T0184653

# Wissenschaftliche Reihe
# Fahrzeugtechnik Universität Stuttgart

**Reihe herausgegeben von**
M. Bargende, Stuttgart, Deutschland
H.-C. Reuss, Stuttgart, Deutschland
J. Wiedemann, Stuttgart, Deutschland

Das Institut für Verbrennungsmotoren und Kraftfahrwesen (IVK) an der Universität Stuttgart erforscht, entwickelt, appliziert und erprobt, in enger Zusammenarbeit mit der Industrie, Elemente bzw. Technologien aus dem Bereich moderner Fahrzeugkonzepte. Das Institut gliedert sich in die drei Bereiche Kraftfahrwesen, Fahrzeugantriebe und Kraftfahrzeug-Mechatronik. Aufgabe dieser Bereiche ist die Ausarbeitung des Themengebietes im Prüfstandsbetrieb, in Theorie und Simulation. Schwerpunkte des Kraftfahrwesens sind hierbei die Aerodynamik, Akustik (NVH), Fahrdynamik und Fahrermodellierung, Leichtbau, Sicherheit, Kraftübertragung sowie Energie und Thermomanagement – auch in Verbindung mit hybriden und batterieelektrischen Fahrzeugkonzepten.

Der Bereich Fahrzeugantriebe widmet sich den Themen Brennverfahrensentwicklung einschließlich Regelungs- und Steuerungskonzeptionen bei zugleich minimierten Emissionen, komplexe Abgasnachbehandlung, Aufladesysteme und -strategien, Hybridsysteme und Betriebsstrategien sowie mechanisch-akustischen Fragestellungen.

Themen der Kraftfahrzeug-Mechatronik sind die Antriebsstrangregelung/Hybride, Elektromobilität, Bordnetz und Energiemanagement, Funktions- und Softwareentwicklung sowie Test und Diagnose.

Die Erfüllung dieser Aufgaben wird prüfstandsseitig neben vielem anderen unterstützt durch 19 Motorenprüfstände, zwei Rollenprüfstände, einen 1:1-Fahrsimulator, einen Antriebsstrangprüfstand, einen Thermowindkanal sowie einen 1:1-Aeroakustikwindkanal.

Die wissenschaftliche Reihe „Fahrzeugtechnik Universität Stuttgart" präsentiert über die am Institut entstandenen Promotionen die hervorragenden Arbeitsergebnisse der Forschungstätigkeiten am IVK.

**Reihe herausgegeben von**

Prof. Dr.-Ing. Michael Bargende
Lehrstuhl Fahrzeugantriebe
Institut für Verbrennungsmotoren und
Kraftfahrwesen, Universität Stuttgart
Stuttgart, Deutschland

Prof. Dr.-Ing. Jochen Wiedemann
Lehrstuhl Kraftfahrwesen
Institut für Verbrennungsmotoren und
Kraftfahrwesen, Universität Stuttgart
Stuttgart, Deutschland

Prof. Dr.-Ing. Hans-Christian Reuss
Lehrstuhl Kraftfahrzeugmechatronik
Institut für Verbrennungsmotoren und
Kraftfahrwesen, Universität Stuttgart
Stuttgart, Deutschland

Weitere Bände in der Reihe http://www.springer.com/series/13535

Florian Winke

# Transient Effects in Simulations of Hybrid Electric Drivetrains

 Springer Vieweg

Florian Winke
Stuttgart, Germany

Dissertation, University of Stuttgart, 2017

D93

Wissenschaftliche Reihe Fahrzeugtechnik Universität Stuttgart
ISBN 978-3-658-22553-7        ISBN 978-3-658-22554-4    (eBook)
https://doi.org/10.1007/978-3-658-22554-4

Library of Congress Control Number: 2018945538

Springer Vieweg
© Springer Fachmedien Wiesbaden GmbH, part of Springer Nature 2019

Printed on acid-free paper

This Springer Vieweg imprint is published by the registered company Springer Fachmedien Wiesbaden GmbH part of Springer Nature
The registered company address is: Abraham-Lincoln-Str. 46, 65189 Wiesbaden, Germany

# Preface

This work was realized during my tenure as a research associate at the Research Institute of Automotive Engineering and Vehicle Engines Stuttgart (FKFS) under the supervision of Prof. Dr.-Ing. M. Bargende. At this point, I want to thank everybody who contributed to the development of this work.

My deep gratitude goes to Prof. Dr.-Ing. M. Bargende for the guidance and support during the past years. I feel incredibly fortunate for the opportunity to work with him.

I want to thank Prof. G. Rizzoni for the suggestions and ideas provided as well as the thoughtful discussions that we had.

I am extremely grateful to all my colleagues from whom I learned a lot, not only from a scientific point of view. I feel honored to have had the opportunity to work in such a diverse and interesting team. Many thanks are also owed to my technical supervisor Dipl.-Ing. Hans-Jürgen Berner for his assistance. Your thoughtful input was always very much appreciated.

Furthermore, I want to thank my family and my girlfriend for the support during this interesting, but also sometimes strenuous time. Without you, this would not have been possible.

Stuttgart                                                                 Florian Winke

# Contents

# Figures

# Tables

# Abbreviations

| | |
|---|---|
| AMDC | ARTEMIS Motorway Driving Cycle |
| AUDC | ARTEMIS Urban Driving Cycle |
| | |
| BDC | Bottom Dead Center |
| BMEP | Brake Mean Effective Pressure |
| BSFC | Brake Specific Fuel Consumption |
| | |
| CFD | Computational Fluid Dynamics |
| $CO_2$ | Carbon Dioxide |
| | |
| ECMS | Equivalent Consumption Minimization Strategy |
| EM | Electric Motor |
| EV | Electric Vehicle |
| | |
| FCEV | Fuel-Cell Electric Vehicle |
| FKFS | Forschungsinstitut für Kraftfahrwesen und Fahrzeugmotoren Stuttgart |
| FMEP | Friction Mean Effective Pressure |
| FRM | Fast Running Model |
| FSM | Full Scale Model |
| FTDC | Firing Top Dead Center |
| | |
| GHG | Greenhouse Gas(es) |
| | |
| HEV | Hybrid Electric Vehicle |
| HR | Hybridization Rate |
| | |
| ICE | Internal Combustion Engine |
| IMEP | Indicated Mean Effective Pressure |
| IVK | Institut für Verbrennungsmotoren und Kraftfahrwesen |

Li-Ion          Lithium-Ion (Battery Chemistry)

MPI             Multi Point Injection
MSS             Mid-Size Sedan
MVM             Mean Value Model

NA              Naturally Aspirated
NEDC            New European Driving Cycle
NiMH            Nickel-Metal Hydride (Battery Chemistry)
NN              Neural Network
NTP             Normal Temperature and Pressure
NVH             Noise Vibration Harshness

PbA             Lead Acid (Battery Chemistry)

RDE             Real Driving Emissions

SCC             Sub-Compact Car
SI              Spark Ignition
SoC             State of Charge
SUV             Sports Utility Vehicle

TC              Turbo Charged
TDC             Top Dead Center

UN              United Nations

WLTC            Worldwide Harmonized Light-Duty Vehicles Test Cycle
WOT             Wide Open Throttle

# Symbols

## Latin Letters

| | | |
|---|---|---|
| $A$ | Area | $m^2$ |
| $C$ | Electric Capacity | F |
| $c$ | Speed of Sound | m/s |
| $E$ | Energy | J |
| $F$ | Force | N |
| $g$ | Gravitational Acceleration | $m/s^2$ |
| $H$ | Enthalpy | J |
| $h$ | Heat Transfer Coefficient | $W/m^2$ |
| $I$ | Electric Current | A |
| $J$ | Cost Function (ECMS) | J |
| $k$ | Thermal Conductivity | W/mK |
| $m$ | Mass | kg |
| $Nu$ | Nusselt Number | - |
| $P$ | Power, Pressure | W, Pa |
| $p$ | Probabilty Factor | - |
| $Pr$ | Prandtl Number | - |
| $Q$ | Heat | J |
| $q$ | Heat Flux | W |
| $R$ | Gear Ratio, Gas Constant | -, J/kgK |
| $r$ | Radius | m |
| $Re$ | Reynolds Number | - |
| $s$ | Equivalence Factor | - |
| $T$ | Torque, Temperature | Nm, K |
| $U$ | Internal Energy | J/kgK |
| $u$ | Torque Split Factor | - |
| $V$ | Voltage, Volume | V, $m^3$ |
| $v$ | Velocity | m/s |
| $W$ | Work | J |
| $w$ | Gas Composition | - |

## Greek Letters

| | | |
|---|---|---|
| $\alpha$ | Thermal Absorptivity | - |
| $\delta$ | Road Slope Angle | ° |
| $\varepsilon$ | Thermal Emissivity | - |
| $\eta$ | Efficiency | - |
| $\kappa$ | Isentropic Heat Capacity | J/kgK |
| $\lambda$ | Air–Fuel Equivalence Ratio | - |
| $\mu$ | Friction Coefficient | - |
| $\omega$ | Rotational Velocity | 1/s |
| $\varphi$ | Crank Angle | °CA |
| $\rho$ | Density | $kg/m^3$ |
| $\sigma$ | Stefan-Boltzmann Constant | J/K |
| $\Theta$ | Rotational Inertia | $kg\ m^2$ |

## Indices

| | |
|---|---|
| 0 | Reference |
| act | Actual |
| aero | Aerodynamic |
| aux | Auxiliary |
| bat | Battery |
| bb | Blow-By |
| cc | Combustion Chamber |
| cha | Charging |
| cl | Clutch |
| d | Drag |
| df | Dynamic Frictional |
| dis | Discharging |
| dyn | Dynamical |
| eff | Effective |
| el | Electrical |
| em | Electric Machine |
| eq | Equivalent |
| f | Frontal, Fuel, Fluid |
| fd | Final Drive |
| fric | Friction |

| | |
|---|---|
| ice | Internal Combustion Engine |
| in | Input |
| ind | Indicated |
| lhv | Lower Heating Value |
| lim | Limit |
| lo | Low |
| mech | Mechanical |
| nom | Nominal |
| norm | Normalized |
| oc | Open Circuit |
| out | Output |
| par | In parallel |
| pwt | Powertrain |
| ref | Reference |
| req | Required |
| roll | Rolling |
| ser | In Series |
| sub | Substitute |
| tot | Total |
| tra | Transmission |
| trac | Traction |
| veh | Vehicle |
| virt | Virtual |
| whe | Wheel |

# Abstract

Due to the increased complexity of hybrid vehicle technology, development of powertrains for hybrid electric vehicles (HEV) can only be performed efficiently by using the best possible simulation technology. Various simulation methods can be used depending on the given boundary conditions and the specific objectives of the investigations. As the dimensioning of drivetrain components has to be addressed in a very early phase of the development process, predictive models are of particular importance.

In general, system and component modeling can be either dynamic or steady-state. While the main advantage of steady-state models is the simplicity and fast computation time, these models are well suited when high-level operation strategies and long-term system behavior need to be evaluated. Also, these models allow for the use of optimization algorithms, like Dynamic Programming, that are not (or only in a restricted way) suitable for dynamic models. Dynamic models provide a more detailed and often physically based approach that leads to more realistic results. The information of the dynamic component behavior is needed for lower-level controls or subsystem design. Subsystem models of different levels of detail are assembled to create a complete system model. The main disadvantages of dynamic models are the increased computation time and the higher complexity compared to steady-state models, leading to higher expenses in terms of time and cost, both for the setup of models and the actual simulations. The selection of a specific modeling approach is therefore always a trade-off with respect to a given use-case.

While the described differences apply for all model types within automotive drivetrains, the greatest differences can usually be observed for models of internal combustion engines (ICE). The long computation time of detailed models of ICE is the reason that even for dynamic drivetrain models the behavior of the ICE is generally modeled by a map-based approach. Common full scale 1D-CFD models of the air path coupled with predictive combustion models provide a realistic representation of ICE behavior but have computation times with real-time factors in the range of 50-100 or more (1 second simulation time takes 50-100 seconds of computation time). Due to their high complexity, the implementation of such models into a complete drivetrain model structure is generally prevented by excessive computation times. However, the real-time factor of 1D-CFD engine models can be significantly decreased by improvements of the air path model of the ICE, while still keeping the general model structure.

This work presents a comparison of different approaches for the simulation of HEV length dynamics with a focus on fuel efficiency. Beginning with the mechanical coupling of the drivetrain, investigations are continued on the component level with dynamic battery and ICE models. Differences of the transient behavior of naturally aspirated and turbocharged engines are investigated and shown, the influence of transient battery behavior is shown within driving cycle simulations.

Detailed 1D-CFD models are compared within an HEV drivetrain to 'traditional' map-based combustion engine models as well as different types of simplified engine models which are able to reduce computing time significantly while keeping the model accuracy at a high level. In a first step, simplified air path models (fast running models) are coupled with a quasi-dimensional, predictive combustion model. In a further step of reducing the computation time, an alternative way of modeling the in cylinder processes are evaluated, by replacing the combustion model with a mean value model. For this approach, the most important influencing factors of the 1D-CFD air path model (temperature, pressure, A/F-ratio) are used as input values into neural networks, while the corresponding outputs are in turn used as feedback for the air path model. However, while the computing speed of the simulation can be further increased, this model type loses the important predictiveness, compared to detailed combustion models.

To cover a wide spectrum of boundary conditions, the performance of the used engine models is evaluated within HEV drivetrain models for vehicles from subcompact class (A-segment) to full-size SUV (J-Segment). Results are shown for both synthetic manoeuvres like accelerations or the New European Driving Cycle, as well as close to real-world driving patterns (Artemis, WLTC). The focus of all investigations is on the trade-off between model quality and computation time.

# Kurzfassung

Aufgrund der erhöhten Komplexität von hybriden Antriebssträngen kann die Entwicklung von Hybridelektrofahrzeugen (HEV) nur dann sinnvoll erfolgen, wenn dabei effiziente Simulationsmethoden zum Einsatz kommen. Verschiedene Ansätze können hier verwendet werden, die sich maßgeblich unterscheiden. Da viele Untersuchungen in einer sehr frühen Phase des Entwicklungszyklus erfolgen müssen, sind prädiktive Modelle von besonderer Bedeutung.

Im Allgemeinen können System- und Komponentenmodelle dynamisch oder quasi-stationär aufgebaut sein. Der Hauptvorteil von quasi-stationären Modellen liegt dabei in ihrer Einfachheit sowie der schnellen Rechenzeit. Deshalb eignen sich solche Modelle insbesondere zur Untersuchung von Betriebsstrategien für das Gesamtsystem sowie Betrachtungen mit langen Zeitskalen. Zusätzlich erlauben solch einfache Modelle oft die Verwendung von erweiterten Optimierungsalgorithmen, wie der Dynamischen Programmierung, die sich nicht (oder nur eingeschränkt) auf dynamische Modelle anwenden lassen. Dynamische Modelle ermöglichen einen zumeist physikalisch basierten Ansatz, der zu realistischeren Ergebnissen führt. Informationen über das dynamische Verhalten werden außerdem meist für die Auslegung und Funktionsentwicklung auf Komponentenebene benötigt. Subsystem Modelle von unterschiedlichem Detaillierungsgrad werden dann zu einem Gesamtsystemmodell zusammengeführt. Der hauptsächliche Nachteil von dynamischen Modellen ist die höhere Komplexität und längere Rechenzeit, was zu einem höheren zeitlichen Aufwand beim Erstellen und Bedaten von Modellen sowie bei der Durchführung von Simulationen führt. Die Auswahl eines spezifischen Modellansatzes ist deshalb immer ein Kompromiss in Bezug auf eine bestimmte Anwendung.

Während diese Unterschiede für alle Arten von Modellen innerhalb automobiler Antriebsstränge gelten, ergeben sich die größten Unterschiede in der Regel für den Verbrennungsmotor (VM). Die langen Rechenzeiten von VM-Modellen führen dazu, dass selbst innerhalb von dynamischen Antriebsstrangmodellen für den VM in der Regel kennfeldbasierte Modelle Anwendung finden. Übliche 1D-Strömungsmodelle des Luftpfades in Kombination mit prädiktiven Ver-

brennungsmodellen erlauben eine realistische Wiedergabe des verbrennungs-
motorischen Verhaltens, führen aber oft zu Rechenzeiten im Bereich 50-100
facher Echtzeit (1 Sekunde Simulationszeit benötigt 50-100 Sekunden Rechenzeit). Aufgrund ihrer hohen Komplexität wird eine Implementierung solcher
Ansätze in Gesamtfahrzeugmodelle in der Regel durch extrem hohe Rechenzeiten verhindert. Die Rechenzeit von 1D-VM Modellen kann allerdings durch
Optimierungen am Strömungsmodell signifikant reduziert werden, während
die generelle Modellstruktur erhalten bleibt.

In dieser Arbeit wird ein detaillierter Vergleich von verschiedenen Ansätzen
zur Längsdynamiksimulation von Hybridfahrzeugen vorgestellt. Der Fokus
liegt dabei stets auf dem Kraftstoffverbrauch. Beginnend mit der mechanischen
Kopplung des Antriebsstrangs führen die Untersuchung weiter auf Komponentenebene, wo Batterie und VM-Modelle mit unterschiedlicher Detaillierung
analysiert werden. Unterschiede die sich dabei zwischen aufgeladenen und
Saugmotoren ergeben werden genauso untersucht wie der Einfluss des tran-
sienten Batterieverhaltens innerhalb von Fahrzyklussimulationen.

Detaillierte 1D-Motormodelle werden innerhalb eines Gesamtsystemmodells
eines Hybridantriebsstrangs mit kennfeldbasierten Modellen verglichen. Außerdem werden vereinfachte VM-Modelle untersucht, die eine deutliche Re-
duktion der Rechenzeit erlauben, während die Simulationsgüte weitestgehend
unbeeinflusst bleibt. In einem ersten Schritt wird dabei lediglich das Strömungsmodell des Luftpfads angepasst, während das quasidimensionale Zylin-
dermodell, und damit die volle Vorhersagefähigkeit, erhalten bleibt. In diesem
Fall spricht man von schnell laufenden Modellen (engl.: Fast Running Models - FRM). In einem weiteren Schritt wird ein alternativer Ansatz zur Ab-
bildung der innermotorischen Vorgänge untersucht, indem das Verbrennungs-
modell durch einen Mittelwert-Ansatz ersetzt wird. In diesem Fall werden die
wichtigsten Einflussgrößen des Strömungsmodells (Temperatur, Druck, Luft-
verhältnis) als Eingangsgrößen für Neuronale Netze verwendet, deren Aus-
gangswerte dann wiederum als Rückkopplung auf das Luftpfadmodell dienen.
Während die Rechenzeit durch solche Anpassungen weiter optimiert werden
kann, verlieren die Modelle, gegenüber detaillierten VM-Modellen, die wich-
tige Vorhersagefähigkeit.

Um ein weites Spektrum von Rahmenbedingungen abzudecken, werden verschiedene Hybridantriebsstränge sowie Fahrzeugkonzepte aus unterschiedlichen Klassen untersucht. Deshalb werden von der Subkompakt Klasse (A-Segment) bis zu Fullsize-SUV (J-Segment) verschiedene Konzepte analysiert. Dabei werden Ergebnisse von synthetischen Manövern wie Beschleunigungskurven oder dem NEFZ genauso betrachtet wie realistischere Fahrkurven (Artemis Zyklen, WLTC). Der Fokus aller Untersuchungen liegt dabei auf dem beschriebenen Kompromiss zwischen Modellgüte und benötigtem Aufwand.

# 1 Introduction

Since the beginning of industrialization, the demand for individual mobility has been continuously increasing. Fast growing populations and people striving to improve their standards of living lead to rapid increases in vehicle numbers and traffic. Figure 1.1 illustrates that the trend towards more vehicles continues until today, even in industrialized regions. Increased purchasing power in developing countries leads to a further growth of vehicle numbers worldwide, as passenger cars continue to evolve from a luxury good towards a commodity. [40]

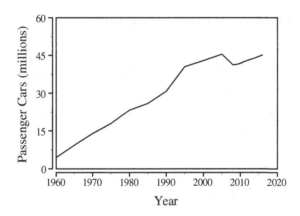

**Figure 1.1:** Development of Registered Passenger Cars in Germany [36]

With the increased usage of vehicles comes increased consumption of fuel and thus higher emissions of greenhouse gases. In the last decades the awareness has been growing that anthropogenic greenhouse gas (GHG) emissions are having an impact on the world's climate. Treaties like the *Kyoto Protocol* [64] and the *Agreement of Paris* [65] document the demand to cut GHG emissions, of which carbon dioxide ($CO_2$) is the most discussed.

© Springer Fachmedien Wiesbaden GmbH, part of Springer Nature 2019
F. Winke, *Transient Effects in Simulations of Hybrid Electric Drivetrains*, Wissenschaftliche Reihe Fahrzeugtechnik Universität Stuttgart, https://doi.org/10.1007/978-3-658-22554-4_1

The fact that the transport sector is a major contributor to the total anthropogenic GHG emissions (compare figure 1.2) leads to the requirement to reduce these emissions.

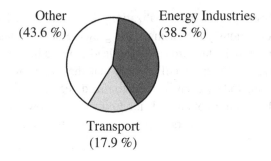

Figure 1.2: GHG Emissions by Sectors in Germany (2014) [63]

Generally, a reduction of $CO_2$ emissions of the transport sector can be realized through different ways: [51]

1. Usage of low carbon fuels or energy from renewable sources

2. Reduction of the annual mileage of vehicles

3. Reduction of driving resistances

4. Optimization of the efficiency of automotive powertrains

A typical example for a low carbon fuel is *Natural Gas* or *Methane* which, due to the higher ratio of hydrogen to carbon atoms, leads to a lower release of $CO_2$ per unit energy compared to conventional fuels. [12] Electricity and Hydrogen are generally $CO_2$-free, but have to be generated from renewable sources. However, this is mostly not the case, so that the greenhouse gas emissions are only emitted at a different location.

The main contribution that automotive development can provide are the reduction of driving resistances and the optimization of the efficiency of powertrains. Hybrid powertrains, which are the main topic of this work, are technically demanding but provide enormous potential to improve powertrain efficiency. [51]

## 1.1 Motivation and Goals of this Work

Due to the higher complexity of hybrid vehicles an efficient analysis and development of such concepts can only be carried out with the help of powerful simulation tools. Crucial development tasks, like concept design or the dimensioning of powertrain components, have to be adressed during early phases of the development process. This is why predictive models are of particular importance. To ensure an efficient use of simulation tools, it is necessary to understand the underlying mechanisms and the influence of models with different complexity on the results of simulations. The assessment of efficient simulation tools is generally a cost to benefit comparison where the quality of the results is compared with the required time to set up and execute models for a specific simulation task. The main goal of this work is to provide a detailed comparison of simulation models with different complexity with regard to this trade off between simulation time and quality of results.

With increasing computational capacity and advanced tools, the importance of simulation within the automotive development process has been rising continuously during the last decades. [39] Sophisticated simulation environments allow the use of detailed, physically based models instead of simplified, abstract algorithms that are mainly based on experimental data or engineering experience. However, especially for the development of complex systems like hybrid powertrains it is necessary to provide models with fast computation times in order to manage the complexity and the number of variants while still enabling reasonable development times. The investigations of this work shall help to evaluate the results of simplified simulations properly to ensure a decision-making process based on valid data.

This work is divided into three main sections. Chapter 1 provides a general introduction into the topic of hybrid powertrains and their simulation as well as the necessary fundamentals for the following sections. Chapters 2 - 4 contain detailed investigations of different submodels, each starting with the basic principles for the specific topic, followed by a description of the main parameters, an overview over the simulations that were carried out and a brief interpretation of the results. Chapter 5 presents the main conclusions of this work and an outlook to possible future works.

## 1.2 Hybrid Powertrains

This section introduces the general fundamentals and typical characteristics of hybrid powertrains as well as the necessary terms and definitions for their analysis. Therefore, a basic overview is given over past and ongoing developments in the field.

The general meaning of the word *hybrid* (from latin *hybrida* or *ibrida*) is described as "anything derived from heterogeneous sources, or composed of elements of different or incongruous kinds". [60] Hybrid approaches are characterized by the fact that multiple solutions are combined for a single feature. Hybrid vehicles are equipped with two or more energy converters as well as two or more energy storage systems that are used for propulsion purposes. [69] The term is most commonly used for vehicles with a powertrain that consists of an internal combustion engine (*ICE*) and one or more electric motors. The terminology hybrid electric vehicle (*HEV*) is well established and helps to differentiate from other systems. Nevertheless, other hybrid propulsion concepts are possible as well and covered by the definition above. Examples are fuel cell electric vehicles (*FCEV*) or hydraulic hybrids that both also contain two energy converters and the corresponding storage systems. [23, 29]

This work is focused solely on HEV that are using an internal combustion engine. Where a distinction is needed, the expression *Non-HEV* [19] is used for conventional vehicles without an electric propulsion unit that are solely powered by an internal combustion engine.

### 1.2.1 History

During the very early stages of the automotive industry, around 1900, multiple propulsion systems were competing for technological and economical predominance. At the time, gasoline engines and electric motors were seen as solutions of almost equal value. The problem with electric vehicles was the same as more than hundred years later, the batteries were too big, too heavy and too expensive, especially when traveling long distances. At the beginning of the twentieth century, a "City Car Concept" was developed. Electric vehicles were intended to be used in urban districts, whereas gasoline powered cars should provide unrestricted mobility in rural areas. As a solution to the arising problem of having to purchase and maintain two vehicles, the concept of the HEV was brought up. [29, 35]

One of the first hybrid vehicles was presented as a concept on the international motor vehicle exhibition in Berlin 1899 by the Belgian company Pieper. The propulsion power for the vehicle was provided by an electric motor, a combustion engine or a combination of the two. An integrated battery pack was intended to be charged through the use of the electric machine as a generator, while the combustion engine had to provide the sum of generator and propulsion power. [35] This concept can be described as a parallel hybrid powertrain (see also section 1.2.2).

On November 23, 1905, Henri Pieper filed a patent for his parallel hybrid at the US Patent and Trademark Office and received the approval on March 2, 1909. The specification of the vehicle already included most of the hybrid functions that are common today, like regeneration of kinetic energy during braking and a simple battery management. [45]

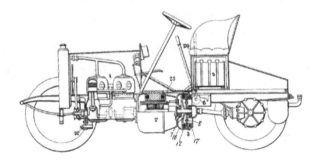

**Figure 1.3:** Illustration out of Henri Pieper's Patent from 1905 [45]

Inspired by the hybrid concept of Pieper, Ferdinand Porsche also developed a hybrid vehicle, the Lohner-Porsche Mixte (at the time, Porsche was working for Ludwig Lohner). The Mixte was based on the Lohner-Porsche electro-mobile that was using wheel hub motors to drive the vehicle. The same motors were used for the Lohner-Porsche Mixte, which can be seen as the first production vehicle with hybrid powertrain. [29] The difference was that the electric power was provided not only by a battery, but also by a generator unit that was powered by a gasoline engine. The generated electric energy could be used to power the wheel hub motors or stored in the battery unit instead. To realize peak power, the electric motors were powered by the battery and generator unit simultaneously. This structure is commonly referred to as series hybrid (see also section 1.2.2). [35]

Despite the promising beginnings, hybrid vehicle concepts were not able to be established against the gasoline powered competition on a significant scale. Later, the self-starter made it easy for all drivers to start gasoline engines and Henry Ford started with the mass production of low-priced, lightweight, gasoline-powered vehicles. Within only a few years, sales of steam and electrically powered vehicles were almost down to zero. One of the last manufacturers of hybrid vehicles, Owen Magnetic, stopped production in 1921. [8]

In the second half of the twentieth century, increasing oil prices and rising awareness of environmental values led to a renaissance of hybrid vehicles. While first prototypes were presented starting in the late 1960s, hybrid vehicles only started to gain significant market shares with the introduction of the Toyota Prius in 1997 in Japan. [18] The first hybrid production vehicle available in Europe was the Audi Duo, based on the Audi A4 Avant, that was launched also in 1997. It combined a Diesel engine on the front axle with an electric motor on the rear axle. However, due to the insufficient demand, production was stopped the following year. [51]

### 1.2.2 Architectures

When an internal combustion engine is combined with one or more electric machines, the components can be arranged in several ways. The determining factors are the number and position of the machines and the couplings between them. This is generally referred to as the powertrain architecture. [68] HEV architectures can be classified into three basic types, parallel, series and power-split, that are discussed in the following. Several special forms of architectures can also be found, but they can all be broken down to one of the three basic architectures or a combination of them.

**Parallel Hybrid**

Parallel hybrids are characterized by the fact that combustion engine and electric motor both have a direct mechanical coupling with the drive wheels. Propulsion power can therefore be provided by one of the machines separately or by a combination of the two so that peak power requirements can be provided by combined operation. The summation of propulsion power is mechanical. Thus, combustion engine and electric machine can be dimensioned relatively small without compromising driving performance. The electric machine can be used as a motor or generator and uses the traction battery as buffer. [12, 43]

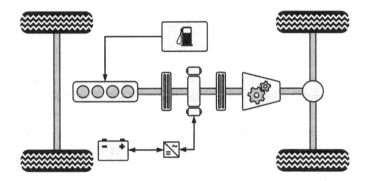

**Figure 1.4:** Parallel Hybrid Architecture

In the simplest case, a parallel hybrid can be realized by implementing a relatively small electric machine (starter-generator) into the belt drive of the combustion engine. This setup is called P1 configuration, which is characterized by a rigid mechanical coupling of the two components. If electric motor and combustion engine can be separated by a clutch, this is referred to as P2 configuration. This is most common for larger electric machines and can be realized either by implementing a separating clutch between the two drive units or by mounting the electric machine directly to the transmission input shaft behind the conventional shifting clutch. P2 hybrids offer the possibility to decouple the combustion engine when driving purely electric so that friction losses can be minimized. This ensures high efficiency for both electric driving and energy regeneration during braking. [29, 49]

An advantage of the parallel hybrid architecture is that only one electric machine is required to realize a relatively simple but effective hybrid system. Due to the direct mechanical coupling of both propulsion units with the drive wheels, conversion losses between mechanical and electrical energy can be kept to a minimum in parallel hybrid powertrains. However, the mechanical coupling can also be a handicap as the fixed relationship of the operation points offers limited flexibility compared to other hybrid architectures. Also, the dynamic requirements for the internal combustion engine are the highest, compared with other hybrid architectures, which is why the focus of this work was set on parallel hybrid powertrains. [49, 69]

**Series Hybrid**

In series hybrids, the combustion engine has no mechanical coupling with the drive wheels. Instead, it drives a generator so that chemical energy (fuel) is converted to electric energy. The summation of power is electrical rather than mechanical so that the drive power is provided completely by one or more electric machines. Due to the decoupling from the drive wheels, this system offers a maximum of flexibility for the operation of the combustion engine. The ICE can be operated within the optimal operation range with minimal dynamic requirements. [12, 43]

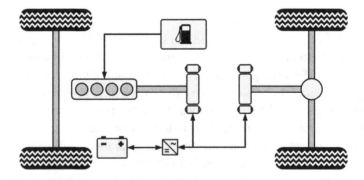

**Figure 1.5:** Series Hybrid Architecture

The series hybrid architecture also ensures high flexibility for the arrangement of components and packaging aspects. Since transmission and shifting clutch may be omitted, this setup is often referred to as *electric transmission* (which is what it would be without the traction battery). As no mechanical connection to a conventional drivetrain is required, wheel hub motors can be used, allowing for optimal manoeuvrability. Functionalities like turning on the spot or moving sideways into narrow parking spaces can be realized. However, the main disadvantage of series hybrids is the high number of energy conversions, which always involve losses. This is why such concepts require a relatively large dimensioning of the propulsion units. Also, in certain conditions, series hybrids have significantly lower efficiency than systems with a mechanical coupling of the engine to the wheels. This applies especially for driving conditions with high power requirements over a long time, like highway driving. [12, 29, 43]

**Powersplit Hybrid**

The structure of powersplit hybrids is characterized by the possibilities to mechanically and electrically merge the power of the propulsion units. For this setup, a power split device (usually a planetary gear set) and two electric machines are required. The use of a conventional transmission is not necessary. In a powersplit hybrid, propulsion power of the engine can only be transferred mechanically to the wheels, when the generator provides a supporting torque so that parts of the mechanical is converted to electrical power. [43, 68]

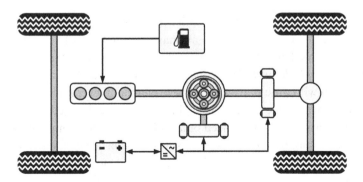

**Figure 1.6:** Powersplit Hybrid Architecture

In contrast to the parallel hybrid, it is not possible to provide the drive power solely by the combustion engine. The torque balance within the planetary gear set always leads to a split of the power into a mechanical and an electrical path. A second electric machine is used to support the combustion engine or regenerate energy while braking. This motor can also be used for all electric driving with the combustion engine turned off. [29]

Due to the continuously variable gear ratio of the planetary gear set, this system allows high flexibility for the operation of the combustion engine, while still providing a mechanical connection to the wheels. As the multiple energy conversion of series hybrids is applied to only a fraction of the power flow, losses in the powertrain can be reduced. Nevertheless, the strengths of this system are utilized especially for low and medium power requirements, while for high power output (highway driving) other configurations are beneficial. [12, 43]

**Other Configurations**

While the architectures described above represent the basic configurations, modified concepts or combinations are possible nevertheless. A *series/parallel hybrid* can be described as a system that allows to change the powertrain configuration from parallel to series and vice versa. It can be realized by the implementation of one or two clutches that are used to switch between the required mechanical connections. The series/parallel architecture is used in many powertrains, for example in the Volkswagen TwinDrive. [11, 43]

Another configuration is that the combustion engine and electric motor are mounted to different axles. As the mechanical coupling between the two is realized by the road between the wheels, this is referred to as *through the road hybrid* or *axle split hybrid*. This concept is a simple way to realize a parallel hybrid system with little impact on the conventional powertrain and high flexibility with regard to the system packaging. Also, it can be used to implement a four-wheel drive without the need of a cardan shaft, for example in compact SUV. The axle split hybrid technology is well established at several car manufacturers, the first production vehicle was the Audi Duo from 1997. [8, 41]

### 1.2.3 Classification

The impact of hybridization on the operation of the powertrain depends not only on the HEV architecture but also to a large extend on the dimensioning of the electric components. Thus, it is helpful to classify HEV with regards to the size of the propulsion units. Conventional (ICE only) and battery electric vehicles appear in figure 1.7 as borderline cases but shall not be discussed in the following, as they are no hybrid vehicles with respect to the definition shown at the beginning of this section.

A common classification of HEV with increasing size of the electric components is the following: [29, 49]

1. Micro Hybrid

2. Mild Hybrid

3. Full Hybrid

4. Plug-In Hybrid

5. Range Extender Electric Vehicle

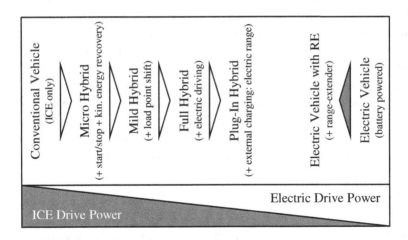

**Figure 1.7:** Spectrum of Hybrid Powertrains

A distinction between the different classifications can be made with respect to several parameters, for example the rated power of the electric machines or the electric power to weight ratio. Another helpful parameter especially for parallel hybrids, is the hybridization rate (HR) [61] that is given by

$$HR = \frac{P_{el}}{P_{tot}} = \frac{P_{el}}{P_{ice} + P_{el}},$$    eq. 1.1

where $P_{el}$ is the output power of the electric machine(s), $P_{ice}$ the rated power of the internal combustion engine and $P_{tot}$ the total drive power of the vehicle. Thus, $HR = 0$ % corresponds to a conventional powertrain, without electric drive power, while $HR = 100$ % implies a battery electric vehicle without internal combustion engine.

**Micro Hybrid**

The simplest way of hybridization at the lowest possible cost is represented by the micro hybrid. Typically, a starter-generator is implemented into the belt drive of the combustion engine, while the conventional powertrain remains unchanged. Therefore, the electric machine has to be compact and also has to work with the conventional boardnet voltage of 12 V, so that the power output is in the range of 4 kW. The hybridization rate is usually below 5 %. [61] The functionalities are limited to an extended start/stop feature with a smoother restart of the combustion engine as well as minor regeneration of brake energy. However, the regenerated energy is only used to reduce boardnet power demand, engine assist functionalities are not possible. The improvements in fuel efficiency are limited to around 5 % and mostly due to the start/stop functionalities so that the biggest improvements are seen for vehicles that spend a significant amount of time waiting at traffic lights or come to stop in traffic. [3, 43]

**Mild Hybrid**

Mild hybrids extend the functionalities of micro hybrids with moderate engine assist capabilities as well as enhanced regenerative braking. Start/stop features are extended so that the engine can be turned off while coasting and braking even at higher vehicle speeds. In some cases, electric driving at very low speeds (*electric creeping*) can be realized. [3] The electric machines used for such systems are typically in the range of 10 kW with an operating voltage of 48 V. [67] The more powerful batteries that are required for such systems also enable other technologies like electric supercharging that allow for a combined optimization of combustion engine and electric components. Fuel efficiency improvements in the range of 10 %, compared to conventional powertrains, can be expected. The order of magnitude of the hybridization rate is 5 - 15 %. [43, 61]

**Full Hybrid**

With a hybridization rate of above 15 % and electric machines of 20 kW or more, [61] full hybrids offer all functionalities described above as well as full electric driving capabilities. The engine can stay turned off at moderate vehicle speeds while the propulsion power is provided by the electric machine and traction battery. Due to the power requirements of the electric components, such systems typically have operating voltages of more than 200 V. [37] Fuel consumption benefits of 15 % or more, compared to conventional powertrains, may be achieved with full hybrids. [12, 52] However, complex energy management strategies are required for the coordination of the hybrid functions while simple, heuristic rules are sufficient for most micro and mild hybrid systems. [49]

**Plug-In Hybrid**

While plug-in hybrids offer all functionalities of full hybrids, their characteristic feature is the ability to recharge the battery by connecting it to an external power source. In most cases this goes along with a larger battery capacity and more electric drive power to ensure extended electric range even for higher power demands. Typically such systems have an operation mode switch to offer the possibility to change from pure electric driving to engine assisted hybrid operation. Also, due to the higher battery capacities, plug-in hybrids allow the use of distinct charge sustaining and charge depleting modes. [27]

The current certification process in Europe favors plug-in hybrids in the way that the electric range is weighted as $CO_2$-free (even though electric power generation is far from being $CO_2$-free for most providers). [20, 21] The most significant downsides of plug-in hybrids are the weight and cost of such systems as they are practically equipped with two full-featured powertrains. [49]

**Range Extender Electric Vehicle**

The term range extender refers to a module that provides electric power to charge the traction battery or to power the electric motor while driving. Such a system may be implemented in electric vehicles to reduce the worries of many customers to get stuck with an empty battery. In the case of concepts with range extenders, there is often confusion whether they should be considered as hybrid or electric vehicles. Some sources categorize them as electric vehicles, [3, 8] which does make sense, as for the typical daily usage the focus is on electric operation. However, range extender electric vehicles conform to the definition for hybrid vehicles at the beginning of this section. In the effort to widely establish electric propulsion systems, both plug-in hybrids and range extender electric vehicles may play a key role in the future. [29]

Apart from conventional piston engines, several different technologies are discussed for the use in range extender units. Examples are rotary engines, Stirling engines or fuel cells, although none of the mentioned got beyond prototype stage yet. The rated power of range extender units varies substantially with typical systems being in the range of 10 - 50 kW. [27]

## 1.3 Control Strategies for Hybrid Vehicles

For hybrid vehicles, the presence of two propulsion units offers an additional degree of freedom. In concepts with only one propulsion system, the mechanical power to drive the vehicle can only be provided by that power source. For conventional vehicles, the power demand of the driver is directly transformed into an actuation of either the combustion engine or the brakes. For hybrids, drive power can be provided by any combination of the two propulsion units. This has to be addressed with the implementation of a control strategy that coordinates the power split between them while taking into account boundary conditions like the battery state of charge (SoC). As the coordination between the components is followed up by a control of the actuators within the single components, this is referred to as *high-level* or *supervisory control*, in contrast to the *low-level* or *component-level control* that are following. Hybrid control strategies are also often referred to as *energy management strategies* because they are mainly controlling the energy flows within the powertrain. Figure 1.8 shows the described two-level control architecture. [43, 58]

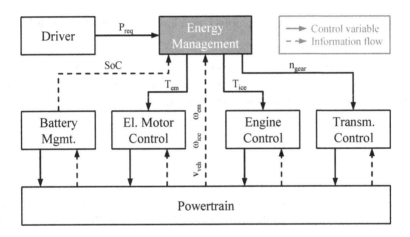

**Figure 1.8:** Typical Control Architecture for HEV [43]

The following section presents an overview of such control strategies for HEV. First, the operation modes that are coordinated by the energy management are described, before possible classifications are introduced. The last part contains a detailed description of the two approaches that were used for this work.

### 1.3.1 Operation Modes

The improvements in fuel economy for HEV compared to conventional vehicles are based on three pillars: [7]

1. *regenerative braking:*
   recovery of kinetic energy during decelerations

2. *electric driving and start/stop:*
   elimination of ICE operation points with low efficiency by switching the engine off

3. *load point shifting:*
   substitution of ICE operation points with low efficiency and such at higher efficiency

While regenerative braking leads to a reduction of the total propulsion energy demand, electric driving, start/stop and load point shifting enable an increase of the mean efficiency of the combustion engine.

### Regenerative Braking

During decelerations of conventional vehicles, all kinetic energy is dissipated to heat in the wheel brakes which means that the energy is lost for the powertrain. Electric machines on the other hand offer the possibility to be operated as generators, so that the kinetic energy can pe partly recovered and stored in the battery. This is referred to as *regenerative braking*. The determining factors for the amount of energy that can possibly be gained with regenerative braking are the vehicle architecture (see section 1.2.2) as well as the size of the electric machine and battery. [49]

As kinetic energy depends directly on the vehicle weight, regenerative braking is especially interesting for heavy vehicles as long as the electric machine and battery are not the limiting factors. Also, driving conditions with frequently changing vehicle speeds (city traffic) provide the highest regeneration potential while this is negligible for near-constant speeds as during highway driving. Modifications of the shifting strategy might be necessary to optimize regeneration efficiency. If the hybrid architecture permits, it is beneficial to decouple the combustion engine during the use of regenerative braking in order to minimize friction losses. Generally, the recovery of brake energy is the major factor for fuel savings of HEV because it directly leads to a reduction of the total propulsion energy demand. [29, 61]

**Electric Driving and Start/Stop**

Electric driving and start/stop can play a major role in reducing friction losses by shutting off the combustion engine whenever it is not required. If the vehicle architecture and the dimensioning of the electric components allow this, the ICE can stay shut off up to moderate vehicle speeds as long as the battery SoC permits. Mild hybrids, that do not have the electrical capacities for extensive electric driving, may benefit substantially from coasting. This means that both propulsion units are operated at zero power, with combustion engine shut off, when the driver actuates neither accelerator nor brake pedal. As for regenerative braking, an optimized shifting strategy may help to increase efficiency due to the different characteristics of electric machines and combustion engines. [41, 48]

The possibility to propel the vehicle silently and (locally) emission-free is of great importance especially, but not only, for plug-in HEV. This may in the future allow the entering of zero emission zones, for example in urban areas with high pollution. [3]

**Load Point Shifting**

For conventional powertrains it is only possible to select the operation point of the combustion engine along the power hyperbola by selecting the most efficient gear. As already mentioned in the introduction of this section, the use of a hybrid powertrain leads to an additional degree of freedom because the power split can also be controlled. For the combustion engine there are basically two parameters that can be addressed with a load point shift, fuel consumption and pollutant emissions. An increase in fuel efficiency can generally be achieved by the use of the electric machine as a generator, resulting in higher load of the combustion engine. Reducing the pollutant emissions with the hybrid operation strategy can be beneficial especially for HEV with a diesel engine. However, it may sometimes be detrimental for the fuel efficiency, resulting in a trade-off between the two optimization goals. [3]

For peak power requirements, an addition of the torque of both propulsion units offers the opportunity to improve the driving dynamics. This operation mode is referred to as *boost*. [29]

**Torque Split Factor**

For the mathematical optimization as well as the analysis of hybrid control strategies, it is convenient to define a variable that directly corresponds with the power split between the two propulsion systems. As the focus of this work was on parallel hybrids, this section introduces a control variable for this architecture. However, the same approach can easily be transferred and extended to series and powersplit hybrid systems. [71] When electric machine and combustion engine are operated at the same speed, the power split between the two is equivalent with the torque split, which leads to the *torque split factor u* [57] that can be expressed as

$$u(t) = \frac{T_{em}(t)}{T_{req}(t)} = \frac{T_{em}(t)}{T_{ice}(t) + T_{em}(t)}. \qquad \text{eq. 1.2}$$

where $T_{req}$ is the torque required at the input of the transmission (controlled by the driver model) and $T_{em}$ and $T_{ice}$ are the torques of the electric machine and internal combustion engine, respectively. The introduced factor regulates the torque distribution as it defines the fraction of the total torque that is provided by the electric machine. For known $T_{req}$ and $u$, the torques of the two propulsion units can be directly calculated as

$$T_{em}(t) = u(t) \cdot T_{req}(t) \qquad \text{eq. 1.3}$$

and

$$T_{ice}(t) = (1 - u(t)) \cdot T_{req}(t). \qquad \text{eq. 1.4}$$

As it is defining the torque distribution, $u$ can directly be correlated with the possible hybrid operation modes as described in the section above. Table 1.1 gives an overview of the operation modes and the corresponding values of $u$.

**Table 1.1:** Operation Modes and Torque Split Factor $u$

| Torque Split Factor | Operation Mode |
|---|---|
| $u = 1$ | El. Driving or Regen. Braking (EM only) |
| $1 > u > 0$ | Boosting |
| $u = 0$ | Conventional Operation (ICE only) |
| $u < 0$ | Battery Charging by Load Point Shifting |

When $u = 1$, all propulsion torque is provided by the electric machine which is the case when the combustion engine is shut off. For positive torques this refers to electric driving, but can also mean regenerative braking, when a negative torque is required. For $u = 0$ all torque is provided by the combustion engine. In that case, the HEV is operated analogous to a conventional vehicle. When $1 > u > 0$, both propulsion devices provide a fraction of the torque demand which is the case for boost operation. For $u < 1$, the electric machine is used in generator operation to increase the load on the combustion engine.

### 1.3.2 Classification

Energy management strategies were used in HEV ever since the first vehicles were developed at the beginning of the 20th century. Since then, several different types of optimization methods have evolved, with the most important categories being *rule-based* and *model-based* strategies. [53, 58]

Rule-based control strategies do not involve explicit optimization tools. Instead, they use a set of rules to determine the controls that should be applied for given boundary conditions. Generally, these rules are based on heuristics, engineering intuition or on information generated in advance with mathematical optimization algorithms. The major advantage of these strategies is their low susceptibility to errors which is why more complex solutions are often preceded with rule-based functions. [43, 53]

Model-based strategies generally use the minimization of a cost function to determine the optimal control solution. They may be further divided into analytical and numerical optimization methods. Ānalytical approaches provide formulations that find the optimal control solution in a closed, analytical form.

One of the most prominent strategies of this type is the *Equivalent Consumption Minimization Strategy* [44] that was used in this work and is further described below. Numerical optimization methods use high computation power to find the global optimal solution for a given driving profile. While these algorithms do not allow a real-time implementation, they provide a valuable design tool. Typical use-cases are benchmark studies and the development of design rules for online strategies. In this work, Dynamic Programming [10] was used as a numeric optimization tool. [43]

Both strategies that were used within this work shall be described in more detail in the following sections.

### 1.3.3 Dynamic Programming

Among the optimization tools that are used to determine the optimal control of hybrid vehicles, *Dynamic Programming* is probably the most used algorithm. It is a numerical method that is capable of providing the optimal control for problems of almost any complexity level. In fact, the complexity of the problem is in most cases only limited by the computational power that is available. However, it is not possible to implement Dynamic Programming in real-time applications. As it requires a priori information about the entire optimization horizon, it is a noncausal algorithm. [9, 10, 43]

The algorithm is based on *Bellman's principle of optimality*: [9]

> An optimal policy has the property that whatever the initial state and initial decision are, the remaining decisions must constitute an optimal policy with regard to the state resulting from the first decision.

This means that the optimal solution for a multistage decision-making problem consists itself of optimal partial solutions for each decision. If a partial solution would not be optimal, meaning it could be substituted by a better one, this would result in a better solution for the problem as a whole. [51]

## General Concept

The concept of the Dynamic Programming algorithm can be explained with a simple example. Here, the problem is to go from point A to B in a minimum of time on a number of one way streets as shown in figure 1.9.

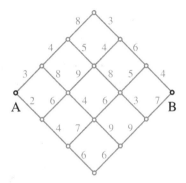

**Figure 1.9:** Travel Time on Map of One-Way Streets

Each Corner can be seen as a *state* while the decision which way to go is the *control policy*. The travel time on each street, shown as the number next to it, is for Dynamic Programming referred to as *arc cost*. In general, this expresses the cost for a transition from one state to another, with respect to the optimization target. [59]

Starting to solve the problem backwards, the decision from C to B (figure 1.10) illustrates the minimization process. For each state, the possible paths at the next step are compared and minimized. The example shows that the fastest way is to go up first, then down (5 + 4 = 9) instead of going down first, then up (3 + 7 = 10). The sum of the arc costs from a specific state to the final solution is referred to as cost-to-go, which can be added up when continuing through the map. Using Bellman's principle, it can be concluded that the optimal solution of the problem as a whole will go the way highlighted with dark arrows, if it includes point C. [59]

**Figure 1.10:** Cost-to-Go from Point C

The procedure shown in figure 1.10 can be repeated for every sub-solution on the map until the cost-to-go is determined for each node.

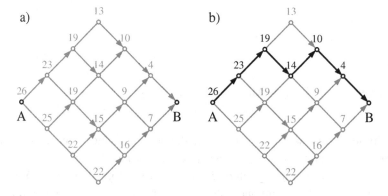

**Figure 1.11:** a) Cost-to-Go and Control Map, b) Finding the Optimal Solution

In addition to the cost-to-go, the control policy (in the figure: the arrows) is also stored while evaluating the single decisions. When the calculation of the cost-to-go map and the corresponding control policy is finished, the result may look as in part a) of figure 1.11. [59]

The minimal time from A to B is already determined as it is equal to the cost-to-go at point A. To find the optimal solution for the problem as a whole, it is possible to start at point A and follow the arrows of the control policy through the map as shown in part b) of figure 1.11. [59]

The main reason why Dynamic Programming is so powerful is that the use of Bellman's principle allows a radical reduction in the required computational power compared to simply calculating every possible trajectory (brute-force approach). For every problem solved with Dynamic Programming, the number of decisions $d$ can be calculated as

$$d = x^2 \cdot (n - 3) + 2 \cdot x \qquad \text{eq. 1.5}$$

where $n$ is the number of steps and $x$ is the number of states at each step. For the same problem there are in total

$$t = x^{n-2} \qquad \text{eq. 1.6}$$

possible trajectories. For a problem of 50 steps with 10 possible states each, Dynamic Programming allows to reduce $10^{48}$ total possible trajectories (eq. 1.6) to only 4270 (eq. 1.5) decisions. [51]

**Application to HEV Energy Management**

The Dynamic Programming algorithm can be used in several different ways to solve the energy management problem in HEV. In this section, the most common way is described, while some possible modifications are shown at the end of the section.

Before the actual optimization process can start, the problem has to be discretized. For this, the driving cycle is split into fixed time steps so that every time step represents one step of the decision-making process. As the state variable, the battery SoC has to be discretized to a state grid. The same has to be done for the control grid which corresponds to the power split between internal combustion engine and electrical system. Thus, the cost corresponds to fuel consumption or in some cases pollutant emissions, depending on the objective of the optimization. Boundary conditions of the powertrain components (SoC limits, maximum torque curves) are considered when setting up the state and control grids, but also during each step of the optimization process. A typical way to exclude operating points that do not satisfy the boundary conditions is to apply near infinite cost so that they will be eliminated during the optimization. Figure 1.12 shows the described approach. [43]

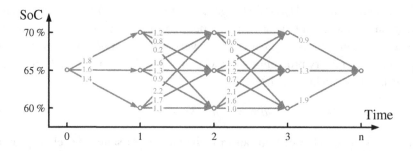

**Figure 1.12:** Application of Dynamic Programming to HEV [43]

The points represent specific states of SoC at the corresponding time step. Each arrow corresponds to a certain value of the power split that would result in the specific change of SoC during a time step. The arc costs on the arrows are the fuel consumption for the given power split during a single time step. For the given system, the optimization process that was described in the section above can be used equally. In practice, it is obvious that values between the discrete grid may occur since variables like the SoC have a continuous state. In this case, the corresponding cost-to-go and control values are based on interpolation. [43]

The problem with the Dynamic Programming algorithm for HEV energy management optimizations is that it can not be implemented as a real-time strategy. As the optimization is evaluated over the complete driving cycle, a priori knowledge about the future boundary conditions is required. Thus, Dynamic Programming was mainly used as a benchmark analysis during this work. Also, it is not possible to use it with dynamic models because the algorithm itself requires a quasi-static modeling of the powertrain. For these purposes, where Dynamic Programming could not be implemented, another strategy was used that is described in the following.

### 1.3.4 Equivalent Consumption Minimization Strategy

The *Equivalent Consumption Minimization Strategy* (ECMS) was developed to provide an algorithm that reduces the global optimization problem over an entire driving profile to an instantaneous minimization that can be solved during the operation based on the actual boundary conditions in the powertrain. The idea behind this strategy is to transform any use of electric energy from the battery to an equivalent fuel consumption that can be directly compared to the actual fuel consumption of the combustion engine. To allow this, a factor is required to make the electric energy equivalent to using (or saving) a certain amount of fuel. This can be used to introduce a cost function $J$ that can be written as [57]

$$J(t,u) = \Delta E_{fuel}(t,u) + s(t) \cdot \Delta E_{el}(t,u), \qquad \text{eq. 1.7}$$

where $\Delta E_{fuel}$ and $\Delta E_{el}$ are the amount of fuel and electric energy that are used in a given time step and $s$ is the equivalence factor which is used to assign an equivalent fuel consumption to a certain amount of electric energy. [44]

For the calculation of the fuel energy used in the combustion engine, the fuel mass flow $\dot{m}_{fuel}$ as well as the corresponding lower heating value $H_{lhv}$ are used. The electric energy can be calculated from the actual current and voltage at the battery. In this case, the open circuit voltage $V_{oc}$ is used so that the battery losses are included for the optimization.

$$\Delta E_{fuel}(t,u) = H_{lhv} \cdot \dot{m}_{fuel}(t,u) \cdot \Delta t \qquad \text{eq. 1.8}$$

$$\Delta E_{el}(t,u) = V_{oc}(t) \cdot I_{bat}(t,u) \cdot \Delta t \qquad \text{eq. 1.9}$$

Knowing these relationships, eq. 1.7 can be divided by the lower heating value of the fuel to calculate the equivalent fuel consumption that gave the ECMS its name: [44]

$$\dot{m}_{eq}(t,u) = \dot{m}_{fuel}(t,u) + \dot{m}_{bat}(t,u) \qquad \text{eq. 1.10}$$

with

$$\dot{m}_{bat}(t,u) = s(t) \cdot \frac{V_{oc}(t) \cdot I_{bat}(t,u)}{H_{lhv}}. \qquad \text{eq. 1.11}$$

It has to be understood that $\dot{m}_{fuel}$ is a *real* fuel consumption, while $\dot{m}_{bat}$ represents a *virtual* consumption that can be positive (battery discharging) or negative (battery charging). It is implemented only for the minimization process and does not correspond to a physical consumption.

The model concept that lays the foundation for this approach was to see the battery as virtual fuel tank. As charge-sustaining hybrids do not use external electrical energy, ultimately all energy that is used to propel the vehicle comes from fuel. Any electrical energy that is used from the battery at a certain time has to be replenished later using fuel from the combustion engine, or through regenerative braking. The energy from regenerative braking can be seen as free virtual fuel in this concept. Any use of energy from the battery that goes beyond this has to be provided by fuel from the engine. The ECMS therefore can be seen as *energy broker*, buying and selling electrical and fuel energy when the actual conditions in the powertrain permit an efficient use. [43, 44]

The evaluation of the equivalence factor $s$ represents the core of the ECMS as this factor is the major influence on the energy management decisions of the strategy. If $s$ is too low, this leads to an excessive use of electrical energy and to a continuously decreasing battery SoC. Too high values of the equivalence factor lead to higher fuel consumption and an overly increasing battery SoC. The equivalence factor is calculated as [57]

$$s(t) = p(t) \cdot s_{dis} + (1 - p(t)) \cdot s_{cha}.$$ eq. 1.12

The parameters $s_{dis}$ and $s_{cha}$ are determined with a variation of the torque split factor over several simulations. For each simulation, a specific, constant value of $u$ is imposed. However, when the required torque is negative, the electric machine always regenerates the maximum amount of energy, no matter what value of $u$ is imposed. The wheel brakes are only used if the electrical system can not provide the full brake torque. The imposed value of $u$ is therefore only applied for positive required torques. During each of these simulations, the energy used in the powertrain is integrated separately for the electrical path and combustion engine. For the electrical path, negative values can occur, as charging of the battery is considered with a negative sign. Each simulation results in a pair of values for $E_{fuel}$ and $E_{el}$ which represent the integrated energies of the combustion engine and electrical path, respectively. These can be plotted as shown in figure 1.13

The case of conventional driving (ICE only), corresponding to $u = 0$, divides the curve in two branches, each with constant gradient. These gradients are equivalent to the parameters $s_{dis}$ and $s_{cha}$. For positive values of the torque split factor, $s_{dis}$ can be determined, while the slope for negative values is used as $s_{cha}$. The boundaries $u_l$ and $u_r$ are given by the upper and lower boundaries of the SoC. The value of $E_{f0}$ determined for $u = 0$ is equal to the energy required to follow the driving cycle without hybrid components. However, the corresponding value of $E_{e0}$ is not zero. It represents the energy stored from regenerative braking reduced by the energy required to start the combustion engine after standstill and to overcome friction losses in the electric path so that the net power output is zero. As no fuel is required to obtain $E_{e0}$ and it is not influenced by the control strategy, this is referred to as *net free energy*. [57]

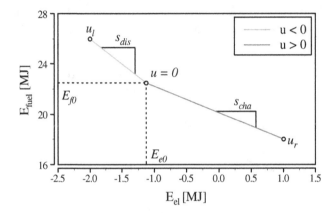

**Figure 1.13:** $E_{fuel}$ and $E_{el}$ for NEDC

It can also be shown that the parameters $s_{dis}$ and $s_{cha}$ can be used to calculate average efficiencies for the conventional and electrical path that can be used for further evaluation: [56]

$$\bar{\eta}_{el} = \sqrt{\frac{s_{cha}}{s_{dis}}} \quad \text{and} \quad \bar{\eta}_{fuel} = \frac{1}{\sqrt{s_{dis} \cdot s_{cha}}} \qquad \text{eq. 1.13}$$

The relationship in eq. 1.12 transfers the core of the ECMS, the evaluation of the equivalence factor $s$, to another problem, the evaluation of the factor $p$ which represents the probability that the electrical energy use at the end of the driving cycle is positive. This relationship can be used as the factors $s_{dis}$ and $s_{cha}$ allow a conversion of used electrical energy into an equivalent fuel consumption (see figure 1.13).

Thus, the optimization comes down to the question if the electrical energy use is positive, quantified by the term $p$, or negative, quantified by the term $1 - p$. The probability $p$ can be calculated as [57]

$$p(t) = \frac{E_{el}^+(t)}{E_{el}^+(t) - E_{el}^-(t)}$$                eq. 1.14

where $E_{el}^+(t)$ is the electrical energy use when constantly driving with $u_r$ for the rest of the cycle and $E_{el}^-(t)$ is the generated energy when constantly driving with $u_l$ for the rest of the cycle. The two values represent the maximum and minimum electrical energy levels that may be reached after the evaluated time horizon. The described quantities are illustrated in figure 1.14. [57]

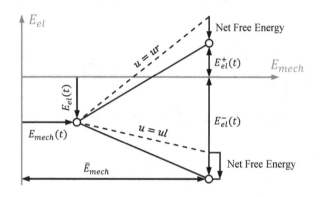

**Figure 1.14:** Variables for the Calculation of the Probability Factor $p$ [57]

With the efficiencies of eq. 1.13 and the parameters $u_l$ and $u_r$ as well as the mechanical energy required for the rest of the evaluated horizon, it is possible to calculate the maximum and minimum electrical energy levels that may be reached at the end of the cycle. However, in a first step it is only possible to evaluate the energy levels for phases with positive propulsion power (dashed lines in figure 1.14). These values have to be corrected by the net free energy, for which an additional parameter $\lambda$ is introduced. It indicates the fraction of the total mechanical energy that can be generated during electric braking: [57]

$$\lambda = \frac{E_{e0}}{\bar{E}_{mech}}.$$                eq. 1.15

This parameter is kept constant for the evaluated horizon and can be used to calculate the net free energy from the remaining mechanical power requirement. Knowing the actual value of $E_{el}(t)$ at a given time, $E_{el}^+(t)$ and $E_{el}^-(t)$ can be calculated as [57]

$$E_{el}^+(t) = E_{el}(t) + \frac{u_r \cdot \left(\overline{E}_{mech} - E_{mech}(t)\right)}{\overline{\eta}_{el}} - \lambda \cdot \left(\overline{E}_{mech} - E_{mech}(t)\right) \qquad \text{eq. 1.16}$$

$$E_{el}^-(t) = E_{el}(t) - \overline{\eta}_{el} \cdot u_l \cdot \left(\overline{E}_{mech} - E_{mech}(t)\right) - \lambda \cdot \left(\overline{E}_{mech} - E_{mech}(t)\right). \qquad \text{eq. 1.17}$$

By combining eq. 1.14, eq. 1.16 and eq. 1.17, the probability factor $p$ can be calculated as

$$p(t) = \frac{\frac{u_r}{\overline{\eta}_{el}} - \lambda}{\frac{u_r}{\overline{\eta}_{el}} + \overline{\eta}_{el} \cdot u_l} + \frac{E_{el}(t)}{\left(\frac{u_r}{\overline{\eta}_{el}} + \overline{\eta}_{el} \cdot u_l\right) \cdot \left(\overline{E}_{mech} - E_{mech}(t)\right)}. \qquad \text{eq. 1.18}$$

This equation allows the calculation of the probability factor $p$, and thus the equivalence factor $s$, for any given time, which is the basis for optimal control of the powertrain. [57]

**Modified Calculation of the Probability Factor**

To avoid the issue of the ECMS that the future driving profile has to be known, a modified calculation of the probability factor $p$ was used in this work. Here, a calculation based solely on the battery state of charge and the corresponding upper and lower limits is used:

$$p(t) = \frac{SoC(t) - SoC_{lim,lo}}{SoC_{lim,up} - SoC_{lim,lo}}. \qquad \text{eq. 1.19}$$

This approach simplifies the calculation procedure and allows more flexibility when changing between different driving profiles. Note that the optimization is still dependent on the input data as the optimal values of $s_{cha}$ and $s_{dis}$ are specific for a particular driving profile. However, this calculation allows to transfer the strategy to models without a-priori knowledge of the driving profile or an online adaption of $s_{cha}$ and $s_{dis}$. It could be shown that the strategy still provides good results. [12]

In some cases, the modified, linear calculation of $p$ may result in problems to stay within the SoC limits. To adress this issue, a parabolic calculation is used close to the limits. The profile of $p$ over the battery state of charge is shown in figure 1.15.

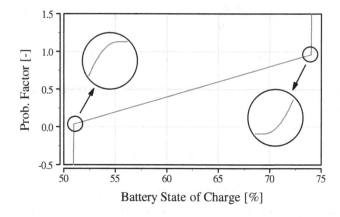

**Figure 1.15:** Probability Factor for Modified Calculation

In addition to this modification, some adaptions were made to account for driveability and avoid excessive switching of the combustion engine operation mode between on and off. For example, a hysteresis was implemented that is triggered when reaching a specific minimum SoC and prohibits operation points that include a discharging of the battery until a certain higher SoC is reached. Also, a function is implemented that only allows a change of the ICE state (on/off) when it is required from the control strategy for a certain amount of time. With this, phases are avoided where the combustion engine would be switched on for very short periods.

These modifications can, in some cases, lead to slightly lower fuel efficiency results than optimization without driveability considerations. However, the resulting control is close to a strategy that may be implemented for production vehicles.

## 1.4 Driving Cycles

Fuel economy and especially the advantages of hybridization (cf. section 1.3.1) depend very much on how a vehicle is driven. Especially the potential to regenerate kinetic energy, but also the benefits from load-point substitution or elimination are affected by the driving profile. As shown in the previous sections, driving resistances are a function of the driving conditions (speed, acceleration, grade) and the vehicle parameters (mass, frontal area, aerodynamic and rolling resistance coefficients). A driving cycle provides the external driving conditions as profiles over time. This section presents an overview of existing driving cycles with a focus on the ones used in this work. In practice, these driving cycles are used for measurements on chassis dynamometers, therefore they are very useful for simulations of the vehicle longitudinal dynamics.

There are two basic types of driving cycles that differ significantly in the way the driving conditions are determined. *Synthetic driving cycles* are defined as abstract speed profiles that allow high reproducibility but do not necessarily represent the conditions during actual road driving. *Real-world driving cycles* instead are based on measurements of vehicle speed and elevation on the road. Therefore, the advantage of real-world driving cycles is that they are very helpful to reproduce driving conditions on the road. While the fuel consumption results of synthetic profiles tend to be optimistic, real-world cycles allow more realistic results. Also, synthetic driving cycles often allow to optimize a vehicle on specific average operation points which is not always improving the fuel efficiency during customer use. [43]

The most popular examples of synthetic driving profiles are the *New European Driving Cycle* (NEDC) [15] and the *Japan 10-15* [33] cycle. A large number of driving cycles can be found for real-world driving cycles, some attempting to reproduce the average driving pattern for a certain region, others with a focus on specific boundary conditions like urban or highway driving. An example for the latter approach was developed within the ARTEMIS project. [1, 2] Measurements of the driving behavior and habits of drivers in Germany, England, France, Greece and Switzerland were evaluated and condensed into a 52 minute long driving cycle. The complete cycle can be divided into three specific profiles that reproduce the typical behavior for urban, extra urban and highway driving. Another example of a real-world driving cycle with high significance is the *Worldwide Harmonized Light-Duty Vehicles Test Cycle* (WLTC). [66] This cycle was developed with the intention to set a global standard for the measurement of fuel consumption and pollutant emissions of light-duty vehicles. The cycle was derived from a total of 765.000 km of data of drivers in Europe, India, the USA, Japan and Korea. Therefore, it gives a realistic reproduction of the actual conditions during road driving. The profile for passenger cars can be divided into four sections with low, medium, high and extra high speeds which are typically used consecutively as a comprehensive driving cycle. [54] Figure 1.16 shows the cycles used for this work.

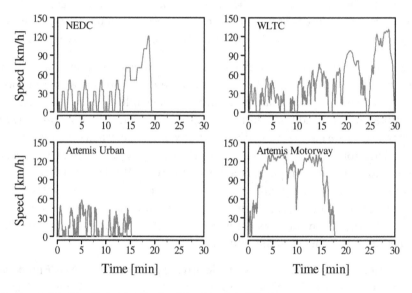

**Figure 1.16:** The Driving Cycles Used in this Work

The driving cycles shown above can be used for different purposes. The NEDC represents a standard cycle and thus provides maximum comparability with other works. However, the dynamic requirements of the NEDC profile are very low, which is why it is not very helpful for the investigation of transient effects. The WLTC as well as the Artemis cycles have more dynamic profiles which is already obvious from the illustrations in figure 1.16. Some of the characteristics of the driving cycles are shown in table 1.2. AUDC and AMDC are the Artemis Urban Driving Cycle and Artemis Motorway Driving Cycle, respectively. Cruising here means driving at a constant speed above zero.

**Table 1.2:** Comparison of the Driving Cycles Used in this Work

| Parameter | NEDC | WLTC | AUDC | AMDC |
|---|---|---|---|---|
| Duration [s] | 1180 | 1800 | 920 | 1068 |
| Distance [km] | 11.0 | 23.3 | 4.5 | 28.7 |
| Avg. Speed [km/h] | 33.6 | 46.5 | 17.5 | 96.9 |
| Max. Speed [km/h] | 120.0 | 131.3 | 57.7 | 131.8 |
| Standing time [%] | 23.7 | 12.6 | 25.8 | 1.3 |
| Pos. acc. time [%] | 20.9 | 43.8 | 35.2 | 42.1 |
| Neg. acc. time [%] | 15.1 | 39.9 | 32.5 | 34.7 |
| Cruising time [%] | 40.3 | 3.7 | 6.5 | 21.9 |
| Max. pos. acc. [m/s$^2$] | 1.04 | 1.67 | 2.86 | 1.9 |
| Avg. pos. acc. [m/s$^2$] | 0.59 | 0.41 | 0.73 | 0.41 |
| Max. neg. acc. [m/s$^2$] | -1.39 | -1.5 | -3.14 | -3.4 |
| Avg. neg. acc. [m/s$^2$] | -0.82 | -0.45 | -0.79 | -0.50 |

It can be seen that the operating conditions vary significantly, not only with respect to average speed, but also standing time and accelerations. While the WLTC follows the approach to represent a wide range of driving characteristics, the Artemis profiles are designed for very specific boundary conditions. Comparing the times spent in the different acceleration states, it is obvious that the large amount of cruising time of the synthetic NEDC can not be found for the real-world cycles. Interestingly, the accelerations within the WLTC are still fairly close to the NEDC for both maximum and average values. For the overall acceleration parameters, the numbers above suggest that the AUDC sets the highest standards with respect to dynamics.

From an energetic point of view, driving resistances (cf. section 2.1.1) can be divided into conservative and nonconservative forces. Since $F_{inertia}$ and $F_{grade}$ are conservative forces, the energy used to overcome these resistances can be gained back when returning to the initial state. For example, $F_{inertia}$ is positive, when the vehicle is accelerated and negative during deceleration. The same can be stated for $F_{grade}$ and the vehicle driving uphill or downhill, but elevation changes shall be neglected here since the presented cycles do not include these. On the other hand, $F_{aero}$ and $F_{roll}$ are frictional, nonconservative forces. Here the energy is simply dissipated which means that it is lost for the powertrain system (thermodynamically, exergy is transformed to anergy).

For HEV, this is of particular interest since these relationships can be used to determine the *regeneration potential*, meaning the share of kinetic and potential energy that can be used to regenerate electrical energy. As the frictional losses due to aerodynamic and rolling resistance act also when the vehicle is decelerating or moving downhill, only part of the conservative forces can be regained. The regeneration potential depends on both the vehicle and the driving conditions. Figure 1.17 illustrates the proportions of the different driving resistances for the used cycles. The values are calculated for a typical midsize sedan (cf. table 2.1). All resistances are normalized to the integrated sum of the positive values which corresponds to the energy that would have to be provided by a conventional powertrain.

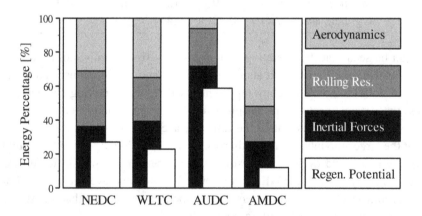

**Figure 1.17:** Energetic Comparison of the Driving Cycles

The illustration shows the size of the impact that the driving conditions have on the energy flows. NEDC and WLTC again show similar characteristics, but a comparison of the two Artemis cycles shows the typical dominance of rolling resistance and inertial forces for low speeds with high dynamics and a shift to the aerodynamics for higher speeds during highway driving. Also the distibution between inertial and frictional forces confirms, what was already stated above, that the AUDC has the biggest influence of dynamic effects. While the distribution between aerodynamic and rolling resistance losses is mainly driven by the average speeds, the inertial forces depend only on changes of the vehicle speed. It is obvious that especially the AUDC is dominated by these inertial effects.

It shall be mentioned that standard driving cycles are not able to reproduce the behavior of all possible drivers but rather a comparison tool during the development of automotive powertrains. The main advantage is the possibility to reproduce certain driving conditions in order to evaluate potentials of new technologies. In practice, each driver has their individual driving style, which is why the measurement of *real driving emissions* (RDE) is gaining importance not only with respect to pollutant emissions but also to determine realistic fuel efficiency values. [54] However, this shall only be a side note here. The results of this work are based on the profiles discussed above.

Driving cycles determine not only the specific driving profile but also the boundary conditions. As some of the driving cycles do not include clear specifications for the speed tolerance, the standards of the NEDC ($\pm 2$ km/h and $\pm 1$ sec.) [15] were used for all cycles. For HEV it is very important to specify the variations of the battery state of charge that are allowed between start and end of the cycle as they may be operated in charge depleting or sustaining mode. As this work focuses on charge sustaining hybrids, it is ensured for all presented results that the battery SoC at the end of the driving cycle lays within $\pm 0.5$ % of the SoC at the start, which can be written as

$$|SoC_{end} - SoC_{start}| \leq 0.5 \ \%. \qquad \text{eq. 1.20}$$

Generally, this is referred to as *SoC-balance*. Comparability between the results can only be guaranteed, if SoC-balance is kept. Technically, it would be more accurate to limit the deviation of the energy content in the battery as the size of the battery has an impact on the results when using the SoC criterium stated above. However, the differences are negligible for the battery sizes investigated in this work so that eq. 1.20 is used for simplicity reasons.

# 2 Powertrain

For this work, multiple powertrain models were developed that differ in the way that dynamic effects are taken into account. All of these models were built up with a modular structure so that it is possible to interchange and compare different sub models of the powertrain components.

The resulting models were used for fuel efficiency investigations on the basis of the driving cycles presented in 1.4. Differences in the dynamic responses and the resulting fuel efficiency were evaluated and compared. This chapter gives an overview of the different approaches that can be used for analyses of vehicle powertrains, before the results of the simulations are presented.

## 2.1 HEV Modeling

This section provides the basics for all models used in this work. More information about the component models that were analyzed specifically is presented in the corresponding chapters and sections.

### 2.1.1 Equations of Motion and Driving Resistances

For most powertrain related investigations, it is sufficient to regard a vehicle as a point mass and its interactions with the surrounding environment. This approach is well suited for simulations of the vehicle longitudinal dynamics and to calculate the power and energy requirements at given boundary conditions, particullarly for a specific speed, acceleration and grade. While lateral dynamics are neglected, analogous to measurements on a chassis dynamometer, this type of simulation can be regarded as one-dimensional analysis. [43]

The equation of motion for the described point mass can be written from the equilibrium of forces acting on the vehicle as shown in figure 2.1:

© Springer Fachmedien Wiesbaden GmbH, part of Springer Nature 2019
F. Winke, *Transient Effects in Simulations of Hybrid Electric Drivetrains*, Wissenschaftliche Reihe Fahrzeugtechnik Universität Stuttgart, https://doi.org/10.1007/978-3-658-22554-4_2

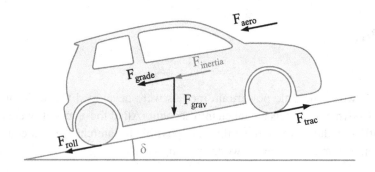

**Figure 2.1:** Forces Acting on a Vehicle

This leads to

$$F_{inertia} = m_{veh} \cdot \frac{dv_{veh}}{dt} = F_{trac} - F_{roll} - F_{aero} - F_{grade},\qquad \text{eq. 2.1}$$

where $F_{inertia}$ is the inertial force, $m_{veh}$ is the effective vehicle mass (including passengers and cargo mass), $v_{veh}$ is the vehicle velocity, $F_{trac}$ is the tractive force as the net sum of powertrain and brakes, $F_{roll}$ is the force due to rolling resistance, $F_{aero}$ is the force due to aerodynamic drag and $F_{grade}$ the force due to road slope. All forces are integrated values for the whole vehicle, which means that rolling resistance and tractive force are representing the sum of all four wheels. The vector arrows in figure 2.1 are only a schematic illustration, the (normalized) vector lengths do not represent a specific scale of the forces.

**Rolling Resistance**

The rolling resistance force is caused due to tire deformation and friction losses within the tires or between tires and road. It can be calculated as the product of a rolling resistance coefficient $c_{roll}$ and the component of the vehicle weight acting perpendicular on the ground:

$$F_{roll} = c_{roll} \cdot m_{veh} \cdot g \cdot \cos \delta,\qquad \text{eq. 2.2}$$

where $g$ is the gravity acceleration and $\delta$ the road slope angle, resulting in $m_{veh} \cdot g \cdot \cos \delta$ as the normal force.

The rolling resistance coefficient $c_{roll}$ is a function of many variables like vehicle speed, tire pressure and temperature as well as road surface conditions. As external factors are complex to measure during road driving and have little to no effect on the chassis dynamometer, $c_{roll}$ is in most cases, and also for this work, modeled as only a function of speed. While the effect of vehicle speed is small at lower values, its influence increases significantly at higher speeds where resonance phenomena become important. [26] To take this into account, a quadratic approach is used here, so that the rolling resistance coefficient becomes

$$c_{roll} = c_{r0} + c_{r1} \cdot v_{veh} + c_{r2} \cdot v_{veh}^2, \qquad \text{eq. 2.3}$$

with $c_{r0}$, $c_{r1}$ and $c_{r2}$ being constant parameters for a given setup. The order of magnitude of $c_{roll}$ is 0.01 or 1 % and can be even lower for optimized tires and tire pressure. [43, 57]

**Aerodynamic Resistance**

Aerodynamic resistance of a vehicle is caused by two effects, viscous friction of air on the vehicle surface and losses due to a pressure difference between the front and the rear of a vehicle, generated by a separation of the air flow. A detailed analysis of the air flow around a vehicle shape is only possible with complex measurements within in a wind tunnel. Thus, the aerodynamic drag is approximated by simplifying the vehicle to a prismatic body with known frontal area $A_f$ and representing the losses with an aerodynamic drag coefficient $c_d$. The actual drag force at a given speed can then be calculated as the product of the stagnation pressure on the frontal area, multiplied by the drag coefficient [26]:

$$F_{aero} = \frac{1}{2} \cdot \rho_{air} \cdot A_f \cdot c_d \cdot v_{veh}^2, \qquad \text{eq. 2.4}$$

where $\rho_{air}$ is the density of the ambient air.

For constant driving on a horizontal road, the total driving resistances are limited to the two described above. Figure 2.2 shows the increase of the driving resistances with speed. The shown values are calculated for a typical midsize sedan. The corresponding parameters can be found in table 2.1.

It can be seen that while rolling resistance is important at low speeds, due to the quadratic increase over speed, aerodynamic drag becomes the dominant force at higher speeds.

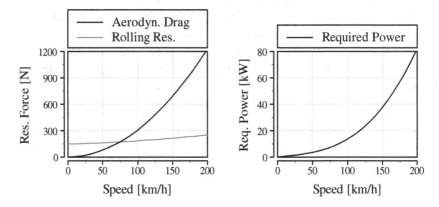

**Figure 2.2:** Resistances for Constant Driving

**Grade Force**

When driving on a non-horizontal road, the vehicle experiences a force induced by gravity. Unlike the aerodynamic and rolling resistances described above, this force is conservative, so that the energy used to climb a slope is not "lost" for the system, but can be gained back when descending. Thus, the net work integrated over a certain path is zero when start and end point are on the same altitude. This is especially important for HEV because of the possibility to electrically regenerate mechanical energy. [26]

The grade force can be calculated as the component of the gravitational force in driving direction by

$$F_{grade} = m_{veh} \cdot g \cdot \sin \delta. \qquad \text{eq. 2.5}$$

**Inertial Forces**

For any given acceleration, a mass causes a fictitious or pseudo force (also called d'Alembert force). [25] For a vehicle point mass, this force was already introduced in eq. 2.1 as $F_{inertia}$. In the case of rotating components, the same has to be taken into account for rotational accelerations. Analogous to the translational part, an equilibrium condition can be obtained as

$$\sum T_o(t) = \Theta \cdot \frac{d}{dt}\omega(t), \qquad \text{eq. 2.6}$$

where $\sum T_o(t)$ is the sum of all outer torques acting on the component at a given time, $\Theta$ is the rotational inertia and $\omega$ is the angular velocity. Generally, this equation is solved in the submodels of the components. However, for simple models (see also chapter 2) it can be convenient to add the inertia of all rotating components to the vehicle mass. [28] Thus, the rotational inertias of all rotating parts are converted to wheel speed and combined to a wheel substitute inertia $\Theta_{whe,sub}$. To do so, the gear ratios within the drivetrain have to be taken into account. For a drivetrain with one differential and a transmission this can be written as

$$\Theta_{whe,sub} = \Theta_{ice} \cdot (R_{tra} \cdot R_{fd})^2 + \Theta_{fd} \cdot R_{fd}^2 + \Theta_{whe}, \qquad \text{eq. 2.7}$$

where $\Theta_{ice}$ is the sum of all inertias rotating with ICE speed, $\Theta_{fd}$ is the sum of all inertias rotating with final drive input speed, $\Theta_{whe}$ is the sum of all inertias rotating with wheel speed and $R_{tra}$ and $R_{fd}$ are the effective gear ratios of the transmission and final drive respectively. It has to be mentioned that to be correct, the mechanical efficiencies of the transmission and final drive would have to be taken into account, which are assumed to be lossless for the equation above. However, as the powertrain inertia is added to the larger vehicle inertia and the described approach is a way to simplify the calculations, the errors of that simplification are generally small and the assumption is acceptable. [26]

When neglecting the wheel slip, vehicle velocity $v_{veh}$ and wheel speed $\omega_{whe}$ can be correlated by using the wheel radius $r_{whe}$:

$$v_{veh} = r_{whe} \cdot \omega_{whe}. \qquad \text{eq. 2.8}$$

Using eq. 2.6, eq. 2.7 and eq. 2.8 and dividing by $r_{whe}$ then leads to the inertial force at the wheel of all rotating parts combined:

$$F_{rot}(t) = \frac{\Theta_{whe,sub}}{r_{whe}^2} \cdot \frac{\mathrm{d}}{\mathrm{d}t} v_{veh}(t).$$                            eq. 2.9

The translational and rotational parts of the inertial forces can then be combined to

$$F_{inertia}(t) = \left( m_{veh} + \frac{\Theta_{whe,sub}}{r_{whe}^2} \right) \cdot \frac{\mathrm{d}}{\mathrm{d}t} v_{veh}(t).$$                            eq. 2.10

As the gear ratio of the transmission appears in the calculation of the wheel substitute inertia (eq. 2.7), it has to be calculated separately for every gear. The quadratic term results in a significant influence of the rotating parts on the vehicle dynamics for high gear ratios, which generally correspond to the lowest gears, and should therefore not be neglected (even for simplified models). [26]

### 2.1.2 Powertrain Modeling Approaches

Two basic approaches are used to simulate the longitudinal dynamics of vehicles. Here, the *kinematic, backward* or *quasi-static* approach can be distinguished from the *dynamic* or *forward* approach. For *kinematic* models the operation points of the propulsion units are calculated directly from a given driving profile using the gear ratios and efficiencies in the drivetrain. The *dynamic* approach reproduces the causality of the powertrain and vehicle in the way that the velocity is calculated based on the operating conditions of the propulsion units that are mostly determined by a torque or power requirement given by a driver model. [26, 43]

### Kinematic Approach

Kinematic simulation models determine the driving conditions at a given time step directly from a given profile. Velocity and elevation are taken from lookup tables, assuming that the vehicle is exactly following the predefined driving profile. With these conditions and taking into account the vehicle parameters, the driving resistances as well as wheel speed and torque can be calculated.

Using the equation of motion of the vehicle (eq. 2.1), the torque at the wheel can be written as

$$T_{pwt} - T_{brake} = F_{trac} \cdot r_{whe} = \left( F_{inertia} + F_{roll} + F_{aero} + F_{grade} \right) \cdot r_{whe}. \quad \text{eq. 2.11}$$

After applying the gear ratios and efficiencies within the powertrain, the operation points of the propulsion units may be determined as shown in figure 2.3. Fuel consumption and electrical power demand for a specific driving cycle can then be calculated with the use of simplified steady-state models like efficiency and consumption maps. [26]

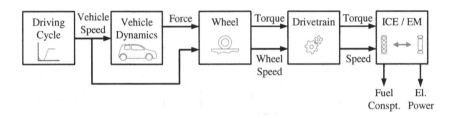

**Figure 2.3:** Kinematic or Backward Simulation Approach (Following [43])

The main advantage of this approach is the simplicity and the fast computation times that can be realized. Maximum consistency between simulations can be guaranteed with kinematic models as it is ensured that the vehicle speed follows the prescribed conditions exactly. Another big advantage is that complex numerical optimization algorithms can be used with kinematic models (Dynamic Programming is one example for such an algorithm). Also, the effort to set up a kinematic model is relatively low. However, unrealistic step responses of speed and torque may occur and different dynamic effects like slipping clutches or engine startup can not be reproduced correctly. [22]

**Dynamic Approach**

The dynamic or forward modeling approach is characterized by a simulation of the vehicle speed based on the actual operating conditions in the powertrain.

The goal is to reproduce the behavior of the components as realistically as possible. For example, transient effects like the engine start up while driving can only be simulated realistically with this approach. The governing differential equation of motion can be written as

$$\frac{\mathrm{d}v_{veh}}{\mathrm{d}t} = \frac{F_{trac} - F_{roll} - F_{aero} - F_{grade}}{m_{veh}}.$$
                                                                                                eq. 2.12

The information flow for a dynamic model is shown in figure 2.4.

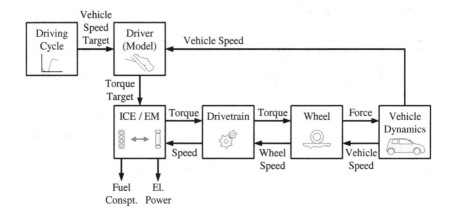

**Figure 2.4:** Dynamic or Forward Simulation Approach (Following [43])

The advantage of this approach is that it allows almost any accuracy level that is required as the sub models can be at any level of detail. For example, as seen in the later chapters, a dynamic powertrain model can be used with a map-based steady-state model of the internal combustion engine. However, a detailed reproduction of the chemical and physical effects within the engine is also possible (and several model types between). As the model approach allows to reproduce the capabilities of the system correctly, it is possible to simulate acceleration profiles according to the powertrain limitations. If the level of detail is appropriate, the development of online control strategies is possible. The main disadvantages are the higher effort that is required to setting up and tune the models, but also higher computation times for the simulations. [62]

Dynamic models also require the implementation of a driver model that controls the torque demand according to the actual and desired vehicle speed. Thus, it is possible to investigate the effect of different types of drivers (cautious/aggressive) on the system behaviour. The downside of the simulation with a driver model is that consistency of the results may be affected since there are always small errors between the target and actual vehicle speed. [43, 62]

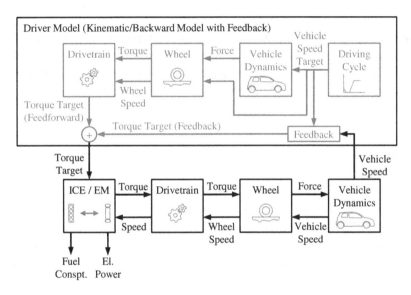

**Figure 2.5:** Backward-Forward Simulation Approach (Following [43])

A possibility to avoid this problem and combine the advantages of a dynamic model with the reproducibility of a kinematic model is the backward-forward simulator illustrated in figure 2.5. Here, the driver model in a dynamic simulator is replaced by a kinematic vehicle model to determine the exact torque that is required to follow the driving profile. With this setup, the vehicle speed will follow the target exactly, unless it is limited by the powertrain capabilities. A feedback term is added to eliminate remaining speed deviations.

The backward-forward simulator offers great functionalities when maximum reproducibility is required while dynamic driving cycles are investigated (as shown in [12]). However, the resulting speed profiles do not reproduce the behavior of a driver on a chassis dynamometer, which is why the backward-forward setup was not used for this work. Instead, the driver model described in section 2.1.3 was used.

### 2.1.3 Powertrain Components

This section provides a brief description of the models for the main powertrain components. Because of the modular structure of the vehicle models used in this work, it is possible to exchange component models of different complexity while leaving the overall model structure unchanged. This was crucial for this work to enable the comparison between the different model types presented in the following chapters.

#### Electric Machine

The electric motor is a permanent magnet synchronous motor modeled by a quasi-static, map-based approach. Due to the good dynamic behavior and the very short response times, a dynamic modeling of the torque build up was not considered necessary. The underlying maps are derived from stationary test bench measurements which are corrected for dynamic operation by using the rotor inertia.

To determine the electrical power consumption for a given operation point, an indicated power is calculated first. This represents the total torque that has to be provided by the electromagnetic system including friction losses. The indicated power $P_{ind}$ can be calculated as

$$P_{ind} = \omega_{em} \cdot \left( T_{em} + T_{fric}(\omega_{em}) \right) \qquad \text{eq. 2.13}$$

where $\omega_{em}$ is the angular velocity of the electric motor, $T_{em}$ is the output torque and $T_{fric}(\omega_{em})$ is the friction torque, which is a function of the angular velocity of the motor.

The indicated power can then be used to calculate the electrical power using the electromechanical efficiency. The latter is a function of the angular velocity of the motor and the output torque and includes losses of the power electronics.

Positive power values are defined to describe the use in motor operation, while negative power values describe generator operation. Depending on the sign of the indicated power, the electrical power $P_{el}$ can be calculated as

$$P_{el} = \begin{cases} P_{ind} \cdot \eta_{em}(\omega_{em}, T_{em}) & \text{for} \quad P_{ind} \leq 0 \quad \text{(generating mode)} \\ P_{ind}/\eta_{em}(\omega_{em}, T_{em}) & \text{for} \quad P_{ind} > 0 \quad \text{(motoring mode)} \end{cases} \qquad \text{eq. 2.14}$$

where $\eta_{em}$ is the electromechanical efficiency. The efficiency map including the losses due to friction and power electronics is shown in figure 2.6.

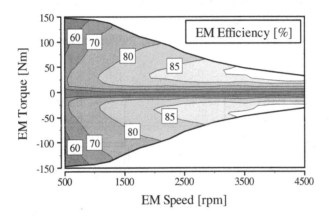

**Figure 2.6:** Electric Motor Efficiency Including Friction and Power Electronics

In a last step, the stationary calculations are corrected with the inertia of the rotor, so that the effective torque $T_{eff}$ can be calculated as

$$T_{eff} = T_{em} - \Theta_{em} \cdot \dot{\omega}_{em} \qquad \text{eq. 2.15}$$

where $\Theta_{em}$ is the rotor inertia and $\dot{\omega}_{em}$ the derivative of the angular velocity with respect to time.

A scaling factor $f_{scale}$ was implemented in the model to allow a variation of the dimensioning of the electric machine. It is defined as

$$f_{scale} = \frac{P_{em,eff}}{P_{em,base}}. \qquad \text{eq. 2.16}$$

This factor is applied to the minimum and maximum torque curves, the friction curve as well as the torque axis of the efficiency map. The inertia of the rotor is also scaled. This factor can be used to easily simulate variations of the hybridization rate and determine an optimal setup of the electric components for given boundary conditions.

To also account for the different torque characteristics of the combustion engines over speed, a virtual ratio $R_{virt}$ can also be applied to adapt the behavior of the electric machine. The effect of the two factors on the maximum torque curve of the motor is illustrated in figure 2.7. The resulting equations are

$$T_{eff} = \frac{T_{basis}}{R_{virt}} \quad \text{and} \quad \omega_{eff} = \omega_{basis} \cdot R_{virt}. \qquad \text{eq. 2.17}$$

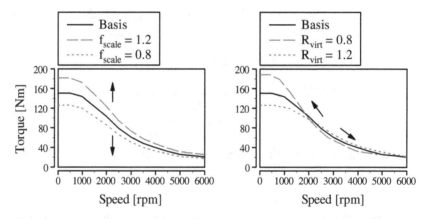

**Figure 2.7:** Scaling of Electric Motor

Note that $f_{scale}$ affects the maximum power and therefore the hybridization rate. It generally also includes a scaling of the available electrical power from the battery by changing the number of cells in parallel. Changes to $R_{virt}$ do not affect the hybridization rate but only the behavior over speed and therefore do not involve adjustments of the battery.

## Internal Combustion Engine

A detailed comparison of different modeling approaches for internal combustion engines was one of the main goals of this work. Thus, a whole chapter is dedicated to this topic so that for the desription of the various engine models, it is referred to chapter 3.

## Battery

The same structure as described above for the internal combustion engine was also used for the battery modeling. Therefore it is referred to chapter 4 for a detailed description of possible modeling approaches as well as the models used in this work.

## Transmission

The transmission model used in this work is based on measurement data of an automated manual transmission. To allow a flexible use, a generic model was developed that can be scaled depending on the maximum input torque and speed. The loss model is based on a polynomial approach that takes into account speed, torque, gear ratio and oil temperature. As thermal investigations were not in the focus of this work, the temperature dependency was not used and steady-state operation temperature assumed instead. Losses of the gearbox and final drive are integrated into a single model, as the test bench data, that was used to calibrate the model, was for gearbox and final drive combined. Figure 2.8 shows loss torque and efficiency for an exemplary gear and a maximum input torque of 250 Nm to illustrate the dependencies.

The total loss torque is applied to the transmission input side, so the input and output torques and speeds are connected by the following equations:

$$T_{out} = (T_{in} - T_{loss}(\omega_{in}, T_{in}, gear)) \cdot R_{gear}, \qquad \text{eq. 2.18}$$

$$\omega_{in} = \omega_{out} \cdot R_{gear}. \qquad \text{eq. 2.19}$$

Generally, transmission losses increase with increasing speed and torque as well as for high transmission ratios. However, at high loads, friction losses decrease towards very low speeds as the oil film between the torque transmitting gears tends to collapse partially or completely for these conditions. This behavior is also visible in figure 2.8.

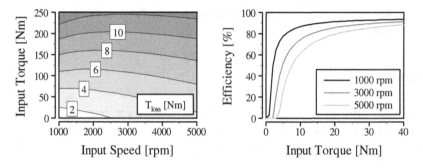

**Figure 2.8:** Transmission Losses and Efficiency for Exemplary Gear

## Clutches

For the models of the dry clutches, three different states are considered. If a clutch is closed, it is modeled as a rigid connection, so that torque and angular velocity are equal on both sides. For an open clutch, input and output side are completely independent and no torque is transmitted. Between these states a slipping clutch acts as a speed converter, with equal torque on input and output side. The transmitted torque $T_{cl}$ can be calculated as

$$T_{cl} = F_{df} \cdot r_{cl}, \qquad \text{eq. 2.20}$$

where $F_{df}$ is the dynamic frictional force and $r_{cl}$ is the effective radius of the clutch. The dynamic frictional force can be calculated as the product of the clutch friction coefficient $\mu_{cl}$ and the actuator force $F_{act}$:

$$F_{df} = \mu_{cl}(F_{act}, \Delta\omega_{cl}) \cdot F_{act}(p_{act}). \qquad \text{eq. 2.21}$$

The clutch friction coefficient $\mu_{cl}$ is a function of the actuator force acting on the clutch surface and the difference of the angular velocities on input and output side. It can be modeled in different complexities or as a first approach held constant. The actuator force can be looked up from a characteristic curve depending on the actuator position $p_{act}$.

**Wheels, Brakes, Tires**

The driving wheels act as the link between the powertrain and the road. At the wheels, the torque of the powertrain is transformed to a tractive force after applying the torque of the wheel brakes. The model of the latter is reduced to a friction torque that is applied depending on the requirements from the vehicle controls. The main equations combine vehicle and wheel speed (already presented in eq. 2.8) as well as tractive force and powertrain and brake torques to

$$v_{veh} = r_{whe} \cdot \omega_{whe},$$ eq. 2.22

$$F_{trac} = \frac{T_{pwt} - T_{brake}}{r_{whe}}.$$ eq. 2.23

A rigid tire model was used in this work, meaning that slipping effects are neglected. This works well, assuming the vehicle is driving on dry asphalt. For low-adherence roads or extreme manoeuvers, a tire slip model should be implemented. [26, 43]

**Auxiliaries**

A number of auxiliary consumers are present in the vehicle that are not specifically modeled but do have an impact on the energy consumption of the vehicle. Examples are cooling water pumps, power steering or the power supply of the control units. Air conditioning is one of the greatest auxiliary consumers but is neglected within this work. As no detailed data was available for most consumers, a simplified approach was used to account for the auxiliaries.

Engine attached accessories like the corresponding cooling water and oil pumps are included in the friction model of the internal combustion engine. All auxiliary consumers that are electrically powered like the control units are combined to a constant electrical load on the traction battery. Considering that all unnecessary consumers are deactivated for fuel economy considerations (air conditioning, lights, audio system all switched off), the boardnet consumption is set to $P_{aux} = 500$ W. [12]

**Driver**

The implemented driver model is based on a PI-controller with speed-dependent parameterization. The driver model outputs a required torque after the gearbox, which makes it independent of the implemented gear shifting strategy. To avoid rapidly changing requirements and ensure a behavior that is comparable to an experienced test bench driver, averaging filters are used. Figure 2.9 shows an example of resulting vehicle speed with given cycle speed and limits.

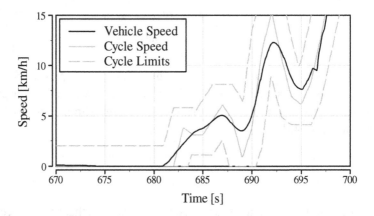

**Figure 2.9:** Vehicle Speed Control by Driver Model

This results in less aggressive torque demand by the driver, with the vehicle speed still staying close to the given cycle profile.

## 2.2 Vehicle Parameters

As the motivation for this work was the understanding of the general mechanisms of the simulation approaches, the investigated vehicle concepts should represent a wide range of possible use cases. In an effort to provide a broad spectrum of boundary conditions, three different vehicle concepts as well as three combustion engines were used in this work. The required data for all component models was taken from the database of FKFS that is derived from finished projects prior to this work. Most of the data is based on test bench measurements from FKFS facilities while some component models are based on data from project partners.

The parameters that determine the vehicle for simulations of the longitudinal dynamics are vehicle mass, drag coefficient, frontal area, and rolling resistance coefficients. Table 2.1 presents the parameter sets for the three vehicle concepts used in this work, which are a subcompact car (SCC), a midsize sedan (MSS) and a sports utility vehicle (SUV). The parameters were selected so that they are within the typical range of each segment. Exemplary production cars would be Mitsubishi i-MiEV or Renault Twingo for SCC, Volkswagen Passat or Honda Accord for MSS as well as Audi Q7 or Volvo XC90 for SUV.

**Table 2.1:** Vehicle Parameters for the Investigated Concepts

|                  | SCC (A-Segment) | MSS (D-Segment) | SUV (J-Segment) |
|------------------|-----------------|-----------------|-----------------|
| Vehicle Mass     | 950 kg          | 1,500 kg        | 2,000 kg        |
| Drag Coefficient | 0.33            | 0.3             | 0.36            |
| Frontal Area     | $2.0 \text{ m}^2$ | $2.2 \text{ m}^2$ | $2.5 \text{ m}^2$ |
| Roll. Res. Coeff.| 0.01            | 0.01            | 0.01            |

The vehicle mass shown in the table above represents the basic weight of the vehicle without driver but also without the electric components for the hybrid propulsion system. When battery and electric machine are scaled (scaling factor introduced in eq. 2.16), the electric systems including casing and cooling circuits are weighted with 8.5 kg/kW, based on approximately 170 kg for the base 20 kW system. The driver is taken into account with 75 kg.

A detailed description of the engine concepts used in this work is presented in chapter 3 so that the explanations here are limited to a quick overview. For the subcompact car, a 0.8 liter naturally aspirated boxer engine was used that was developed within this projects to meet the specific boundary conditions for an urban HEV in an effort to make a reasonable comparison to purely electrically driven vehicles (EV). [72, 73] Two other engines are used for the midsize sedan and SUV interchangeably, one of them naturally aspirated and one turbo charged. With displacements of 1.5 liter for the turbo charged engine and 2.0 liter for the naturally aspirated engine, both provide 123 kW of effective power. All concepts are SI engines and were developed through extensive simulation work. Table 2.2 presents an overview of the engine concepts.

**Table 2.2:** Main Characteristics of the Used ICE Concepts

| Indication | 08NA | 20NA | 15TC |
|---|---|---|---|
| Displacement | 774 cc | 1,999 cc | 1,498 cc |
| Number of Cylinders | 2 | 4 | 4 |
| Configuration | Flat (Boxer) | Inline | Inline |
| Injection | Manifold | Direct | Direct |
| Rated Torque | 75 Nm | 197 Nm | 238 Nm |
| @ Speed | 4,500 rpm | 4,500 rpm | 1,500 rpm |
| Rated Power | 38 kW | 123 kW | 123 kW |
| @ Speed | 5,000 rpm | 6,500 rpm | 5,000 rpm |

The size of the electric motor is varied to a wide extent within this work with the help of the scaling factor and virtual ratio presented in eq. 2.16 and eq. 2.17. The basis for all concepts is a permanent-magnet synchronous motor with a rated power of 20 kW and a maximum torque of 139 Nm. The maximum speed is 6,000 rpm.

While the Subcompact concept with the 2-cylinder 08NA engine is combined with a 5-gear transmission, the other engine and vehicle combinations are equipped with 7 gears. Both models are based on measurement data of automated manual transmissions. The transmission ratios are shown in table 2.3.

**Table 2.3:** Transmission Ratios

|  | 5-Gear | 7-Gear |
|---|---|---|
| Ratio Gear 1 | 3.727 | 3.1875 |
| Ratio Gear 2 | 2.136 | 2.1905 |
| Ratio Gear 3 | 1.414 | 1.5172 |
| Ratio Gear 4 | 1.121 | 1.0571 |
| Ratio Gear 5 | 0.892 | 0.7381 |
| Ratio Gear 6 | - | 0.5574 |
| Ratio Gear 7 | - | 0.4328 |

The ratio of the final drive depends on the specific combinations of engine and vehicle concept. The corresponding optimization will be presented in the following section.

## 2.3 Simulations

This section provides an overview of the results of the simulations with respect to the powertrain dynamics of the simulation models. All components were modeled quasi-statically to ensure that the differences that are obtained are solely from the powertrain model structure described above.

### 2.3.1 Final Drive Optimization

In order to adapt the transmission ratios to the characteristics of the investigated engine concepts, the final drive ratios were optimized for each concept individually. To provide comparability between the concepts, the final drive ratio was optimized for maximum velocity in $5^{th}$ gear so that the speed characteristics of the different engine types come to effect. As the electrical system can only be used for a very limited amount of time and thus is only of low importance for highway driving, the investigations were made for ICE only operation. Figure 2.10 illustrates an exemplary acceleration.

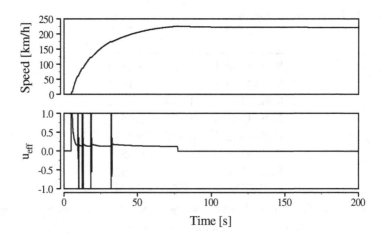

**Figure 2.10:** Exemplary Acceleration to $v_{max}$ for MSS with 15TC

The electric machine was used during the acceleration process before switching to ICE only mode. The mode switch from electric assist to ICE only mode can be seen in the effective torque split factor at 75 seconds.

A variation of the final drive ratio leads to a distinct optimum of the maximum speed, as illustrated in figure 2.11.

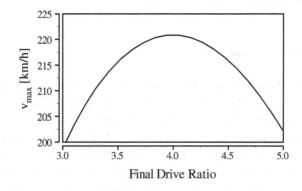

**Figure 2.11:** Final Drive Variation for MSS with 15TC

The same optimization process was evaluated for all used combinations of engine and vehicle concept. The results are shown in table 2.4. As the turbo charged 15TC unfolds its power already at lower speeds, compared to the 20NA concept (cf. section 3.2), the optimization leads to substantially lower final drive ratios and thus lower engine speeds for the same gear. Because of the overall higher driving resistances, the resulting ratios for the SUV are higher than for the midsize sedan and the same engine. The values determined here were used for all following simulations.

**Table 2.4:** Optimized Final Drive Ratios

|                        | 08NA | 20NA | 15TC |
|------------------------|------|------|------|
| Subcompact Vehicle     | 4.65 | -    | -    |
| Midsize Sedan          | -    | 5.02 | 3.92 |
| Sports Utility Vehicle | -    | 6.17 | 4.78 |

While the hybridization ratio is varied widely within some of the following simulations, the final drive ratio remains untouched as it is only depending on the combustion engine. Also, the speed characteristic of the electric machine is only adapted to fit the speed characteristic of the corresponding engine, and not by a change of the output power (cf. section 2.1.3).

### 2.3.2 Model Comparison

This section provides a comparison of the different powertrain modeling approaches. As described in section 2.1.2, the individual model types differ substantially in the way transient effects are considered. Therefore, different dynamic responses can occur depending on the boundary conditions of the selected driving cycle. To give a closer understanding of the underlying mechanisms, the first part of this sections focuses on a detailed investigation of single drivetrain simulations. After this, an overview of multiple vehicle and engine combinations is given.

Figure 2.12 shows a comparison of dynamic and kinematic simulations of the SCC concept with a Hybridization rate of HR = 0.3 in the NEDC.

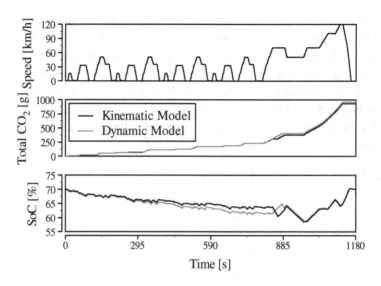

**Figure 2.12:** Dyn. and Kin. Models in NEDC (SCC - HR = 0.3)

It can be seen that the differences concerning the $CO_2$ emissions is relatively low, which is due to the low dynamic requirements of this synthetic driving cycle. When isolating the behavior of the $CO_2$ emissions during the urban part, the patterns are technically identical. However, a comparison of the battery SoC shows that the main differences occur during the urban part of the driving cycle that contains more transient fractions. The identical $CO_2$ emissions at the end of the urban part can only be held through increased usage of the electric machine. The lower SoC then has to be compensated during the extra urban cycle. It can be seen that the differences mainly occur in phases when the ICE is switched on and off frequently as during the urban part of the NEDC.

An analysis of the more dynamic AUDC then leads to increased differences between the models. The results are illustrated in figure 2.13. As the different phases of the driving cycle can not be separated as clearly as for the NEDC, the results show a steady drift of the $CO_2$ emissions between the two model types. Also, it can be seen that due to the high variations in the speed profile and the rapid changes between accelerations and decelerations, the SoC profile is relatively flat. Thus, the electric energy storage could be relatively small here.

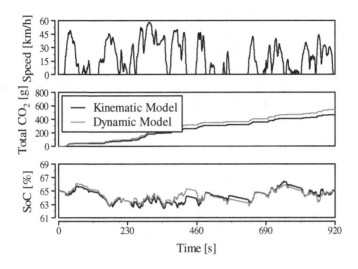

**Figure 2.13:** Dyn. and Kin. Models in AUDC (SCC - HR = 0.3)

For an energy-efficient layout of the powertrain, it is necessary adjust the electric components (electric motor, battery, power electronics) so that they make an optimal match with the selected ICE. For a given combustion engine, such an optimization is possible through a variation of the hybridization rate as shown in figure 2.14.

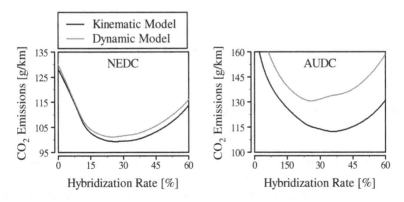

**Figure 2.14:** Dynamic and Kinematic Models - HR-Variations - SCC, 08NA

The investigations show that kinematic simulations may deliver acceptable results for the NEDC that contains long periods of constant cruising. However, when changing to a driving cycle with distinct transient requirements, the limitations of the kinematic approach become obvious very quickly. For an analysis of real-world driving as in the AUDC, dynamic simulation models should be used. Figure 2.15 gives an overview of some other combinations of vehicle and engine concepts. The tendencies described above can be confirmed from these results.

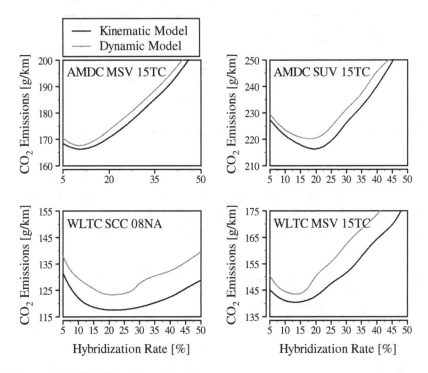

**Figure 2.15:** HR-Variations Overview

Some general conclusions that can be drawn from the findings above shall also be described briefly. First, it can be seen that for the motorway driving cycle AMDC, the optimum of the HR variation is found for very low HR values. This is the case because for driving at high speeds with relatively low transient requirements the ICE can be operated at good efficiencies so that the hybrid functionalities are only needed to a very limited extent. However, for the case of the SUV concept, the weight of the vehicle is so high that a large electric machine can be used to efficiently recuperate kinetic energy even with limited speed variations, leading to higher HR values compared to the MSV. The WLTC results are very similar to what can be observed in the NEDC, which is not surprising when considering the similarities within the energetic analysis presented in chapter 1.4.

## 2.4 Interpretation of the Results

The results presented in this chapter show that the choice of a certain modeling approach has a significant impact on the $CO_2$ emissions calculated during the driving cycle simulations. Also it could be shown that the required modeling depth can be depending on the boundary conditions of a given speed profile. Therefore, the selection of an efficient simulation tool has to be carried out application-specific.

However, with differences between the model types for some cases up to almost 20 %, it is clear that the quasi-static approach reaches its limits rather quickly. Especially when simulating realistic driving patterns, kinematic models generally tend to produce very optimistic results.

As the additional effort for the setup of dynamic powertrain models compared to kinematic models is reasonable with most modern simulation environments, the general recommendation is to work with dynamic simulation models wherever it is possible. Without the use of sophisticated component models, the computation times are still relatively low so that in most cases there are no major barriers preventing this step.

# 3 Internal Combustion Engine

Within a parallel HEV powertrain, the internal combustion engine represents the determining component with respect to fuel consumption and pollutant emissions. Several different approaches can be taken for the simulation of internal combustion engines. The following list presents the most common model types in order of increasing complexity. [43]

1. Map-Based Models

2. Map-Based Models with Lumped Parameter Dynamics

3. Mean Value Models

4. 1D-CFD Models

5. 3D-CFD Models

While numbers 1. to 3. represent simplified approaches, simulations of the *Computational Fluid Dynamics (CFD)* are a very powerful tool, enabling a physics-based evaluation of the processes inside internal combustion engines. However, the highly versatile functionalities of such models comes at the cost of high computation times as illustrated in figure 3.1.

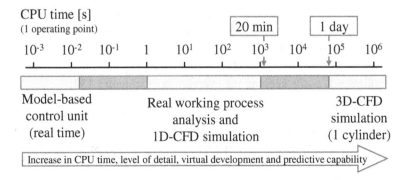

**Figure 3.1:** Approximate CPU time for Different Model Types [6]

© Springer Fachmedien Wiesbaden GmbH, part of Springer Nature 2019
F. Winke, *Transient Effects in Simulations of Hybrid Electric Drivetrains*, Wissenschaftliche Reihe Fahrzeugtechnik Universität Stuttgart, https://doi.org/10.1007/978-3-658-22554-4_3

Map-based models assume the ICE to be a perfect actuator with infinitely short response times. Typically, the engine is characterized with a maximum torque curve and a consumption or efficiency map. In some cases a friction curve (or map) is added to correctly represent the towing operation and run down. Engine speed is taken from the overall powertrain model and torque from the operation strategy (HEV) or driver model (non-HEV). The effective consumption or efficiency can then be interpolated directly from the map. A simple way to include dynamic behavior in the torque response of the engine is to include some kind of lumped-parameter model that limits the maximum torque and possibly adapts the consumption from the map accordingly. As map-based models are often sufficient for energy management and system analysis and also allow very fast computation, this model type represents the most common approach for vehicle simulations. [26]

In contrast to the abstract modeling approaches mentioned above, CFD simulations describe the general physical behavior of the working medium inside the air path and combustion chamber. 3D-CFD models provide the possibility to realistically simulate flow fields of almost any complexity. 1D-CFD models are especially useful for the simulation of flow in pipes, thus offering a great tool to analyze the intake and exhaust track. Mean Value Models present an effort to keep some of the physics-based structure of flow simulations but with a simplified approach that does not take account of the cyclic operation of combustion engines. Therefore, most functionalities of full scale CFD models are lost. Increasing computational capacities have lead to a wide spectrum of uses for both 1D and 3D models. However, the computation times of 3D-CFD simulations are still too high to enable a coupling with powertrain models. For 1D-CFD models this coupling is possible with high, but manageable computation times. Also, adaptions to 1D-CFD models can help to reduce computation time significantly while keeping the overall model structure. [39]

This chapter presents a detailed comparison of different combustion engine models within a hybrid electric powertrain. First, the governing equations and the fundamentals of combustion engine modeling are presented that are required to understand the differences between the model types. After a brief description of the engine used for the presentend investigations, the simulation results of the engine and vehicle simulations are shown.

## 3.1 Internal Combustion Engine Modeling

### 3.1.1 Thermodynamics

From a thermodynamic point of view, the combustion chamber can be classified as open, dynamic system in which all properties are highly variable with respect to time and location. The major state and process variables can be classified as shown in figure 3.2.

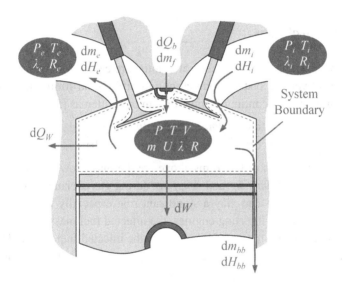

**Figure 3.2:** Thermodynamic System Combustion Chamber [5]

Each working cycle consists of a number of thermodynamic processes which can be categorized as follows: [46]

- *mass transfer:*
  while the intake and exhaust valves are opened, mass flows $dm_i$ and $dm_e$ can transfer the system boundaries; also, the blow-by mass $dm_{bb}$ and for direct injecting engines the fuel mass $dm_f$ have to be taken into account.

- *energy transfer:*
  the heat $dQ_b$ is released by the chemical process of combustion; the work $dW$ as well as the heat $dQ_w$ is transferred from the working gas across the system boundaries; all mass transfers have to be included with the corresponding enthalpy flows.

- *change of internal energy:*
  all of the above mentioned mass and energy transfers can lead to a change of the internal energy $dU$ within the system.

**Conservation of Mass**

Variations of mass within the control volume are calculated by the sum of all mass flows over the system boundaries. In differential form with respect to crank angle $d\varphi$, this continuity equation can be written as [5]

$$\frac{dm}{d\varphi} = \frac{dm_i}{d\varphi} + \frac{dm_e}{d\varphi} + \frac{dm_{bb}}{d\varphi} + \frac{dm_f}{d\varphi}. \qquad \text{eq. 3.1}$$

The major proportion of the terms above are the mass flows through the inlet and exhaust valve, $dm_i/d\varphi$ and $dm_e/d\varphi$ during gas exchange. The blow-by mass flow $dm_{bb}/d\varphi$ can play a significant role especially for high cylinder pressures. For direct injecting engines, the injected fuel mass $dm_f/d\varphi$ has to be considered, while for manifold injection the injected fuel is included in the intake mass flow so that $dm_f/d\varphi = 0$ during the whole working cycle. [39]

**Conservation of Energy**

The first law of thermodynamics for an open, dynamic system forms the basic for all further calculations. In differential form with respect to crank angle $\varphi$ it can be written as [5]

$$\frac{dU}{d\varphi} = \frac{dQ_b}{d\varphi} + \frac{dQ_w}{d\varphi} + \frac{dH_i}{d\varphi} + \frac{dH_e}{d\varphi} + \frac{dH_{bb}}{d\varphi} + \frac{dW}{d\varphi}. \qquad \text{eq. 3.2}$$

Here, $dU/d\varphi$ is the change of internal energy in the system. The burn rate $dQ_b/d\varphi$ represents the heat of the combustion process differentiated with respect to $\varphi$. Together with the wall heat transfer rate $dQ_w/d\varphi$ it results in the heat release rate. For simulations of the working process, the burn rate $dQ_b/d\varphi$ is modeled and calculated based on various boundary conditions. The enthalpy flows $dH_i/d\varphi$, $dH_e/d\varphi$ and $dH_{bb}/d\varphi$ are due to the corresponding mass flows through the intake and exhaust valves as well as the blow-by mass flow. Finally, $dW/d\varphi$ corresponds to the pressure-volume work done at the piston surface so that $dW/d\varphi = P \cdot dV/d\varphi$. [39]

**Equation of State**

The thermodynamic equation of state can generally be written as [5]

$$PV = mR_iT, \qquad\qquad \text{eq. 3.3}$$

where $P$ is pressure, $V$ volume, $m$ mass, $R_i$ is the specific gas constant and $T$ temperature. It is important to note that the specific gas constant is variable for different boundary conditions as it is a function of the gas composition. Dissociation processes also have an effect on the specific gas constant as the concentrations of single species are affected, so that $R_i = f(P,T,w)$, where $w$ is the gas composition. [24] In differential form with respect to the crank angle $\varphi$, the thermodynamic equation of state can be written as [5]

$$P\frac{dV}{d\varphi} + V\frac{dP}{d\varphi} = mR_i\frac{dT}{d\varphi} + mT\frac{dR_i}{d\varphi} + R_iT\frac{dm}{d\varphi}. \qquad \text{eq. 3.4}$$

**Thermodynamic Properties**

For simulations of all processes in the combustion chamber, correct calculations of the thermodynamic properties of the working gas are crucial. While more than 1,000 different chemical species can be observed during engine combustion processes, simplifications have to be made. [70] All common modeling approaches are focused on 9 - 20 species with the greatest impact on the overall gas properties while the other species are omitted due to their negligible impact. As most reactions between the chemical species take place very quickly, especially for high temperatures, the assumption can be made that the main species are in chemical equilibrium. With this assumption, the properties of the mixture can be calculated from the species. [24]

Since the first polynomial approaches were presented in the 1930s, various other concepts have been published that allow a very accurate calculation of the thermodynamic properties. For more information on the topic, it is referred to specialized works. [32, 34, 75] Generally, all thermodynamic properties are a function of pressure, temperature and the gas composition. [24]

At this point it shall be mentioned that for all investigations and simulations in this work, the ambient conditions where assumed to be at Normal Temperature and Pressure (NTP). Where it is necessary, reference altitude was selected to be sea level, so that Temperature and Pressure are $T_{ref} = 20\ °C = 293.15\ K$ and $p_{ref} = 1.013\ bar$ and the density of air is $\rho_{air} = 1.2041\ kg/m^3$. [17, 31]

### 3.1.2 Heat Transfer

Thermodynamically, heat transfer characterizes exchange of energy between thermodynamic systems and the environment surrounding them. There are three fundamental modes of heat transfer, conduction, convection and radiation. For all of these modes, the energy flow occurs always from higher to lower temperature. [31]

### Conduction

Conduction occurs within a substance itself or with another substance that is placed in physical contact. Every temperature gradient results in an exchange of thermal energy through conduction. It can be described by *Fourier's Law of Conduction* as

$$\vec{q} = -k \cdot \nabla T = -k \cdot \left( \frac{\partial T}{\partial x} + \frac{\partial T}{\partial y} + \frac{\partial T}{\partial z} \right), \qquad \text{eq. 3.5}$$

where $\vec{q}$ is the local heat flux density, $k$ the substance's *conductivity*, which represents a material constant and $\nabla T$ is the local temperature gradient. The latter can be obtained as the partial derivatives of temperature with respect to all dimensions in space.

Conduction represents the most important mode of heat transfer for rigid bodies. The conductive heat loss $\dot{Q}$ through the wall of a pipe (for example in the exhaust system) can be written as

$$\dot{Q} = 2\pi \cdot k \cdot l \cdot \frac{T_i - T_o}{\ln\left(\frac{r_o}{r_i}\right)}, \qquad \text{eq. 3.6}$$

where $l$ is the length of the pipe, $T_i$ and $T_o$ are the temperatures on the inside and outside, respectively, and $r_o$ and $r_i$ are the outside and inside radius of the pipe, respectively.

**Convection**

Convective heat transfer is characterized by a transfer of matter that carries part of the thermal energy and thus influences the overall heat transfer. As it is based on the movement of molecules within a certain substance, convection does not occur in rigid bodies but only in fluids.

There are two types of convection, depending on the main cause for the transfer of matter. *Forced convection* is characterized by an imposed movement of the fluid from external surface forces which is generally induced by a fan or pump. This is the dominating mechanism of heat between any gaseous medium inside the airpath of combustion engines and the surrounding structure. For *natural* or *free convection*, the movement of the molecules is induced by gradients in the density of the fluid. These gradients results from the temperature differences in the fluid and result in less dense components rising and more dense components sinking. Therefore, the main driver for natural convection is the gravitational field.

For both convection modes, *Newton's Law of Cooling* can be used as

$$\dot{Q} = h \cdot A \cdot (T_s - T_f), \qquad \text{eq. 3.7}$$

where $h$ is the heat transfer coefficient, $A$ is the Area that is exposed to convection and $T_s$ and $T_f$ are the temperatures of the surface and fluid, respectively. The heat transfer coefficient $h$ can be calculated using a set of dimensionless parameters that characterize the actual conditions inside the fluid. The most important parameter here is the Nusselt number $Nu$ which describes the ratio of the total and conductive heat transfer. Thus, a Nusselt number of $Nu \approx 1$ describes conditions with practically no motion in the fluid so that conduction is dominant. Large Nusselt numbers typically describe turbulent flow conditions with strong influence of molecular movement. The thermal boundary layer can be characterized with the help of other dimensionless parameters like the Prandtl number $Pr$ which describes the characteristic thermal properties of a fluid, or the Grashof number $Gr$ that approximates the ratio of buoyancy and viscous forces. The Reynolds number $Re$ characterizes the (imposed) flow conditions so that the Nusselt number can generally be written as

$$Nu = f(Re, Pr, \delta_L),$$ eq. 3.8

for forced convection and

$$Nu = f(Gr, Pr, \delta_L),$$ eq. 3.9

for free convection, where $\delta_L$ is the characteristic length that describes the scale of the system.

## Radiation

Thermal energy transfer through electromagnetic waves is called *radiation*. Every substance with a temperature above 0 Kelvin emits energy through radiation. Unlike conduction and convection, no physical contact between two substances or any medium between is required for radiation to occur. The absorption and emission of radiative energy are characterized by the absorptivity $\alpha$ and emissivity $\varepsilon$ of a substance. *Kirchhoff's Law of Thermal Radiation* links these two parameters to

$$\varepsilon = \alpha.$$ eq. 3.10

The radiative heat transfer from a substance to the environment can be calculated as

$$\dot{Q} = \varepsilon \cdot \sigma \cdot A \cdot (T_s^4 - T_e^4),$$ eq. 3.11

where $\sigma$ is the Stefan-Boltzmann constant, $A$ the surface area and $T_s$ and $T_e$ are the temperatures of the surface and the environment, respectively.

## In-Cylinder Wall Heat Transfer

A correct specification of the in-cylinder wall heat transfer is crucial for the simulation of internal combustion engines. As the overall heat transfer between working medium and the engine structure is always a combination of the three modes of heat transfer described above, and the conditions in a combustion chamber are highly unsteady, empirical models can be used. A number of approaches are available [4, 30, 74] that are based on Newton's Law shown in eq. 3.7, so that the wall heat transfer $\dot{Q}_w$ can be calculated as

$$\dot{Q}_w = h_w \cdot A_{cc} \cdot (T_g - T_w), \qquad \text{eq. 3.12}$$

where $A_{cc}$ is the surface of the combustion chamber and $T_g$ and $T_w$ are the temperatures of the working gas and cylinder wall, respectively. The differences of the models are in the calculation of the heat transfer coefficient $h$ which generally is a function of the mean gas temperature $T_m$, the in-cylinder pressure $P$, engine speed $\omega$, the cylinder volume $V$, as well as a combustion related term $\Delta$,

$$h_w = f(T_m, P, \omega, V, \Delta). \qquad \text{eq. 3.13}$$

The models of Bargende [4] and Woschni [74] were used for this work.

### 3.1.3 Computational Fluid Dynamics

The understanding and management of the flow field inside the air path of internal combustion engines is crucial for their efficient development. However, analytical solutions can only be provided for simple flow problems like straight cylinders or the flow between parallel plates. Complex flow phenomena as in real combustion engines can only be analyzed by experiments or numerical calculations. The general approach for the solution of flow problems with numerical tools can be divided into four steps. [38, 42]

1. Definition of the Problem

2. Mathematical Formulation
   (Systems of Differential Equations)

3. Discretization
   (Systems of Algebraic Equations)

4. Numeric Solution
   (Iterative Methods)

In many cases, a reduction of the spatial dimensions is possible, depending on the characteristic of the problem. The common model types for engine simulations are 3D, 1D and 0D models. Time is similarly discretized into a sequence of small intervals, which will be described at the end of this section.

**3D-CFD**

The fundamental equations for the solution of flow problems of any complexity are given by the Navier-Stokes-Equations which are defined by the conservation equations for mass, energy and momentum in differential form. In some sources, the term Navier-Stokes-Equations is used only for the momentum conservation equations. However, the full set of conservation equations is always required to solve flow problems. For 3D-CFD simulations, the fluid domain is divided into a finite number of volumes (cells) that form a mesh of up to several millions of elements. This mesh is used as framework for the solution of the governing equations. Figure 3.3 illustrates the concept of a finite volume approach were state variables are calculated for each cell and transfer variables for each boundary to other cells. [14]

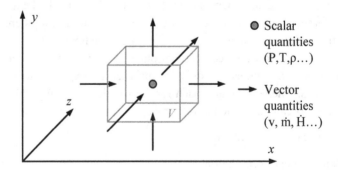

**Figure 3.3:** Three Dimensional Discretization

Using the local velocity vector $\vec{u}$, the gravitation vector $\vec{g}$ and the Nabla operator $\vec{\nabla}$

$$\vec{u} = \begin{bmatrix} u_x \\ u_y \\ u_z \end{bmatrix}, \quad \vec{g} = \begin{bmatrix} g_x \\ g_y \\ g_z \end{bmatrix}, \quad \vec{\nabla} = \begin{bmatrix} \frac{\partial}{\partial x} \\ \frac{\partial}{\partial y} \\ \frac{\partial}{\partial z} \end{bmatrix}, \qquad \text{eq. 3.14}$$

as well as the identity $I$ and tension matrix $\tau$

$$I = \begin{bmatrix} 1 & 0 & 0 \\ 0 & 1 & 0 \\ 0 & 0 & 1 \end{bmatrix}, \quad \tau = \begin{bmatrix} \tau_{xx} & \tau_{xy} & \tau_{xz} \\ \tau_{yx} & \tau_{yy} & \tau_{yz} \\ \tau_{zx} & \tau_{zy} & \tau_{zz} \end{bmatrix}, \qquad \text{eq. 3.15}$$

the conservation equations for mass, energy and momentum can be written as [38]

$$\frac{\partial}{\partial t}\rho + \vec{\nabla} \cdot (\rho \cdot \vec{u}) = 0, \qquad \text{eq. 3.16}$$

$$\frac{\partial}{\partial t}\left(\rho \cdot \left(e + \frac{1}{2} \cdot \vec{u}^2\right)\right) + \vec{\nabla} \cdot \left(\rho \cdot \vec{u} \cdot \left(h + \frac{1}{2} \cdot \vec{u}^2\right) - \tau \cdot \vec{u} - \lambda \cdot \vec{\nabla}T\right) \qquad \text{eq. 3.17}$$
$$= \rho \cdot \vec{g} \cdot \vec{u} + \rho \cdot \dot{q}_R,$$

$$\frac{\partial}{\partial t}(\rho \cdot \vec{u}) + \vec{\nabla} \cdot (\rho \cdot \vec{u} \times \vec{u} + P \cdot I - \tau) = \rho \cdot \vec{g}. \qquad \text{eq. 3.18}$$

If a mixture of several species is used as working medium, the mass conservation equation has to be solved for each species individually and potentially extended by a production and sink term to account for chemical reactions.

**1D-CFD**

The one dimensional or 1D-CFD simulation approach reduces the flow field to a single flow stream, neglecting the variations in the cross section of the flow. The values of all relevant quantities are thus assumed to be constant in the cross section. The resulting discretization is illustrated in figure 3.4.

This approach sufficiently satisfies the demands of most pipe flow problems. Many pipe flow situations that physically can not be modeled with 1D simulations can be represented with simplified models. Examples are friction losses at abrupt changes in the cross section of the flow, where turbulent regions with backflow may be present that can not be modeled correctly in basic 1D models. However, the use of correction functions with the help of nondimensional parameters help to avoid these limitations so that the overall flow behavior can be modeled at very high accuracy. [39]

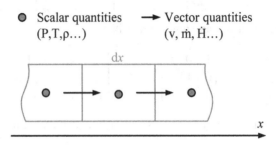

**Figure 3.4:** One Dimensional Discretization

While the general Navier-Stokes equations shown above are still valid for one dimensional flow, they may be written in a simpler form. The continuity equation for 1D-CFD can be written in volumetric terms as

$$\frac{\partial}{\partial t}\rho + \frac{\partial}{\partial x}(\rho \cdot u) = 0, \qquad \text{eq. 3.19}$$

or specifically with respect to mass as

$$\frac{d}{dt}m = \dot{m}_1 - \dot{m}_2 = \rho_1 \cdot A_1 \cdot u_1 - \rho_2 \cdot A_2 \cdot u_2. \qquad \text{eq. 3.20}$$

For the flow simulations in internal combustion engines, the potential energy variations due to elevation changes can generally be omitted. The energy conservation equation can then be used with respect to the internal energy as

$$\frac{d}{dt}E = -P \cdot \frac{d}{dt}V + \dot{m}_1 \cdot H_1 - \dot{m}_2 \cdot H_2 - h \cdot A_{wall} \cdot (T_{fluid} - T_{wall}), \qquad \text{eq. 3.21}$$

where $H_1$ and $H_2$ are the total specific enthalpies at the boundaries, $h$ is the heat transfer coefficient between flow and pipe wall, $A_{wall}$ is the surface area of the pipe exposed to the flow and $T_{fluid}$ and $T_{wall}$ are the temperatures of the fluid and pipe wall, respectively.

The momentum equation for one dimensional flow becomes

$$\frac{\partial}{\partial t}(\rho \cdot u) + \frac{\partial}{\partial x}(\rho \cdot u^2 + P - \tau) = \rho \cdot g_x. \qquad \text{eq. 3.22}$$

or specifically with respect to momentum (again neglecting the effect of gravitational forces)

$$\frac{d}{dt}(m \cdot u) = P_1 \cdot A_1 - P_2 \cdot A_2 + \dot{m}_1 \cdot u_1 - \dot{m}_2 \cdot u_2 - \sum F_{loss}, \qquad \text{eq. 3.23}$$

where the indices 1 and 2 stand for the specific boundaries and $\sum F_{loss}$ represents the sum of all frictional and pressure losses that are typically calculated using table data for certain flow patterns (bends, tapers, restrictions).

**0D and Quasi-Dimenional Models**

In 0D models, spatial variations in the fluid domain are neglected so that averaged values are considered for the quantities of interest. Typically the combustion chamber is modeled with such models when simulating the air path with 1D-CFD. The complex flow field inside the combustion chamber can not be modeled with 0D approaches, but averaged quantities of pressure and temperature can be simulated with very good accuracy.

A common approach to enhance the quality of 0D models is to divide the combustion chamber into two separate zones for burned and unburned mixture during the combustion phase. While the pressure is generally assumed to be constant for both zones, the temperature, gas composition and thermodynamic properties are calculated individually for each zone, enabling more detailed calculations and for example the simulation of pollutant emission formation.

Quasi-dimensional models are characterized by the fact that some geometrical dependencies are taken into account in 0D models. The quality of this approach can not replace the functionalities of 3D models, but they can ensure a certain amount of predictability while keeping the computation times relatively low.

### 3.1.4  Time Discretization

In order to ensure numerical stability, the time step in CFD simulations can not be selected randomly high, but must be restricted to satisfy the Courant condition (also known as Courant-Friedrichs-Lewy or CFL condition): [16]

$$\Delta t < \frac{\Delta x}{|u| + c},$$                    eq. 3.24

where $\Delta t$ is the time step, $u$ the fluid velocity, $c$ is the speed of sound and $\Delta x$ the smallest discretized element length (distance between two grid points). The practical interpretation of this condition is that the time step in the simulation must be lower than the time it takes a (shock) wave to travel from one grid point to the next. If the time step is selected larger than this value, the information propagated by the wave can not be evaluated correctly which leads to numerical instability and incorrect results.

An analysis of eq. 3.24 shows that the main limitation to the time step is the discretization length $\Delta x$, where high values should be sought. This is not always possible as the simulation quality depends highly on the model discretization, but it should always be selected with attention to the computation time. The other factors are the fluid velocity and speed of sound where high values lead to low time steps. The speed of sound can be calculated as

$$c = \sqrt{\kappa \cdot R \cdot T},$$                    eq. 3.25

where $\kappa$ is the isentropic heat capacity ratio, $R$ is the specific gas constant and $T$ is the temperature. This shows that a well-chosen discretization length is especially important in areas with high fluid velocity and high temperatures. Therefore, the critical components limiting the time step for engine simulations are typically in the exhaust ports, where high temperatures, high fluid velocities and short dicretization length come together.

The discretization of flow systems has a very high impact on the quality and computation times of simulation models. A detailed analysis of the influence is shown later in this chapter.

### 3.1.5 Engine Friction

The Schwarzmeier-Reulein friction model was used to specify the friction during all simulations with detailed engine models. The model was originally developed for diesel engines, but also allows good modeling of SI engine friction, especially for highly turbocharged engines. It allows an easy calibration as only one reference point is required which usually is enough to reach a good match of the actual friction losses in the engine. [50, 55]

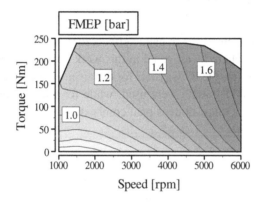

**Figure 3.5:** Exemplary Friction Mean Effective Pressure Map for 15TC

The original Schwarzmeier model includes separate terms for auxiliary loads like fuel or coolant pumps, which for this work were included in the overall friction losses and not modeled individually. The model takes into account oil temperature and brake mean effective pressure which both are related to the reference operation point. Multipliers for each term allow an adaption of the model for each specific case:

$$
\begin{aligned}
FMEP = FMEP_{ref} \\
+ c1 \cdot \left( \frac{c_m}{T_{Wall}^{1.68}} - \frac{c_{m,ref}}{T_{Wall,ref}^{1.68}} \right) + c2 \cdot \left( \frac{BMEP}{T_{Wall}^{1.68}} - \frac{BMEP_{m,ref}}{T_{Wall,ref}^{1.68}} \right) \\
+ c3 \cdot \left( \frac{(d \cdot n)^2}{T_{Oil}^{1.49}} - \frac{(d \cdot n_{ref})^2}{T_{Oil,ref}^{1.49}} \right) + c4 \cdot \left( \frac{BMEP}{T_{Oil}^{1.49}} - \frac{BMEP_{ref}}{T_{Oil,ref}^{1.49}} \right),
\end{aligned}
\quad \text{eq. 3.26}
$$

where $c_m$ is the mean piston speed, $T_{Wall}$ the cylinder wall temperature, $BMEP$ is brake mean effective pressure and $T_{Oil}$ the engine oil temperature. $c_m$ is linked directly to the engine speed and $T_{Wall}$ is calculated using the models and equations described above. $T_{Oil}$ is in this work taken from maps with respect to engine speed and load.

## 3.2  Used Engine Models

To provide an overview of the dynamic effects for several engine types, three different concepts where investigated in this work which are described in the following. To optimize the ICE layouts for the given boundary conditions, 1D-CFD Simulations (GT-Power) were used, while for the simulation of the in-cylinder processes, a quasi-dimensional combustion model was implemented (FKFS user-cylinder).

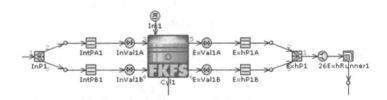

**Figure 3.6:** Detail of GT-Suite Flow Model with FKFS User-Cylinder

### 3.2.1  0.8 Liter Naturally Aspirated Engine

For the subcompact vehicle concept (cf. section 2.2) that was optimized as an urban hybrid, a relatively small, naturally aspirated engine was selected. The goal was to create a compact, simple and cheap engine concept for the HEV drivetrain. A combined power of 60 kW for ICE and EM was required, so that the powertrain optimization lead to a power requirement for the ICE of slightly below 40 kW.

A detailed evaluation of the powertrain and engine optimization process was already published in [71, 72]. Table 3.1 presents the main parameters of this engine concept. The 08NA will also be used to illustrate the process of creating the simplified engine models in sections 3.2.4 to 3.2.6.

**Table 3.1:** Main Characteristics of 08NA Concept

| Indication | 08NA |
| --- | --- |
| Displacement | 774 cc |
| Configuration | Flat 2 (Boxer) |
| Injection | Manifold |
| Rated Torque | 75 Nm |
| Rated Power | 38 kW |

The decision was made for a naturally aspirated 2-cylinder SI engine, leading to a displaced volume of 774 cc for the given power requirements. To also take into account the NVH behavior of a small engine operated almost exclusively at high loads, a boxer setup was chosen (flat engine). In order to keep the ICE setup simple, it was decided to go with manifold injection (MPI) and not include variabilities in the valve train (no cam phasing). The air-fuel ratio is kept at stoichiometric conditions over the complete operating range with no full load enrichment. Figure 3.7 illustrates the power curve and specific fuel consumption map for this engine concept.

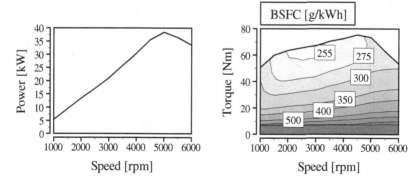

**Figure 3.7:** Power Curve and Fuel Comsumption Map for 08NA

### 3.2.2 2.0 Liter Naturally Aspirated Engine

This engine concept was developed with the intention to provide a typical engine concept for a midsize sedan. Also, a comparison between a turbocharged and naturally aspirated engine of equivalent power was intended so that the 20NA concept and the 15TC (described below) have the same rated power of 123 kW.

**Table 3.2:** Main Characteristics of 20NA Concept

| Indication | 20NA |
|---|---|
| Displacement | 1,999 cc |
| Configuration | Inline 4 |
| Injection | Direct |
| Rated Torque | 197 Nm |
| Rated Power | 123 kW |

The power requirement was defined in advance as 150 kW for the combination of ICE and electric machine. Vehicle optimization simulations as shown in chapter 2 lead to a distribution of 123 kW for the ICE and 27 kW for the electric machine. Both the 20NA and 15TC are inline 4 cylinder concepts with direct injection. Also, both concepts include camshaft phasing on the inlet and outlet sides, representing state-of-the-art engine technology.

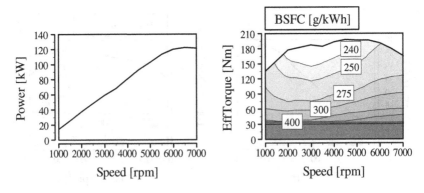

**Figure 3.8:** Power Curve and Fuel Comsumption Map for 20NA

### 3.2.3 1.5 Liter Turbo Charged Engine

As described in the section above, this concept was used as a state-of-the-art engine for a typical midsize sedan. The characteristics of the turbocharged engine are especially interesting because of the distinct transient behavior (turbo-lag). The main parmeters of the 15TC engine are listed in table 3.3.

**Table 3.3:** Main Characteristics of 15TC Concept

| Indication | 15TC |
|---|---|
| Displacement | 1,498 cc |
| Configuration | Inline 4 |
| Injection | Direct |
| Rated Torque | 238 Nm |
| Rated Power | 123 kW |

Compared to the 20NA concept with equivalent power, the down-sized 15TC engine unfolds its power at significantly lower speed, thus provides higher torque. The consumption and power characteristics are illustrated in figure 3.9. It is important to mention that the 15TC engine is also operated at stoichiometric conditions throughout the whole operation range. Instead of mixture enrichment, the maximum torque curve is limited so that operation points with excessive turbine inlet temperatures are avoided.

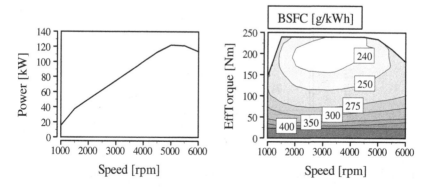

**Figure 3.9:** Power Curve and Fuel Comsumption Map for 15TC

### 3.2.4 Fast Running Engine Models

To illustrate the process of developing the simplified models, the 08NA is used exemplarily in this and the following section. All engine specific results shown in this and the following section refer to this engine concept.

The conventional 1D-CFD models of the described ICE with a real-time factor in the range of 100 (the exact computation time always depends on the simulated operation point and boundary conditions) serve as a basis for the further investigations presented in the following chapter and will be referred to as full scale model (FSM). The real-time factor describes the relationship between simulated time and required computation time, thus providing a measurement for the speed of a simulation. For a real-time factor of 100, 1 second simulation time takes 100 seconds of computation time.

In a first step of increasing the computation speed of the basic model, the air path was simplified to speed up the CFD calculations while remaining the cylinder model unchanged. As the general structure of the CFD model is kept the same and only the flow geometry is modified with the goal of increasing the maximum possible time steps, this is referred to as fast running model (FRM). The advantage of this approach is that the model functionalities remain untouched with only low losses in model accuracy. Figure 3.10 exemplarily shows the correlation of model accuracy and computation speed.

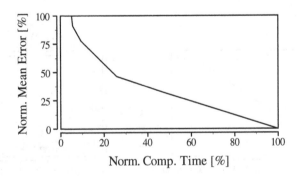

**Figure 3.10:** Development of Mean Error and Computation Time

The condition in eq. 3.24 (Courant condition) showed that the maximum possible time step is proportional to the selected discretization length, when the boundary conditions remain constant. This leads to the conclusion that for every FRM a trade-off between model accuracy and computation speed has to be made. When beginning the FRM development process, big improvements in computation time can be made with only minor cutbacks in model accuracy. With ongoing process the improvements in computation time become smaller while the model accuracy decreases further. The decision when to stop the simplification process depends on the model properties and the application.

For the application in this study, the model was intended to keep a good level of accuracy. As a measure of the model accuracy, the WOT torque curve was used, which had to be kept within a margin of 1 % error compared to the basic model. Even with this high level of accuracy, the computation time could be reduced by up to 95 %, leading to a real-time factor of the FRM of approximately 5. A comparison of the maximum torque curves is illustrated in figure 3.11.

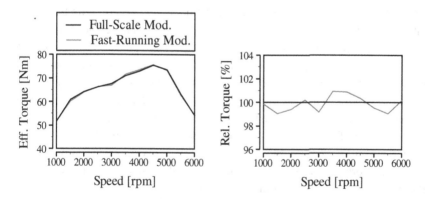

**Figure 3.11:** Model Comparison of Torque at WOT

Figure 3.12 illustrates the pressure profiles at one inlet and one outlet valve. It can be seen that the pressure profiles at the inlet and outlet valves also produce a very good match. The comparison of the pressure after the exhaust valve shows clearly that some high-frequency pulsations are lost due to the enhanced discretization and the larger time steps. However, the characteristics of the profile are reproduced very well.

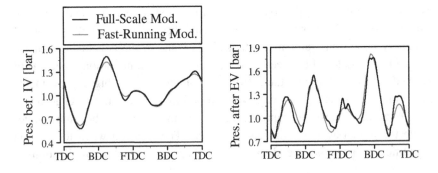

**Figure 3.12:** FRM Pressure Profiles Comparison (5000 rpm WOT)

### 3.2.5 Mean Value Models

In a further step of simplifying the engine model and speeding up the computation time, the quasi-dimensional cylinder models were replaced by a mean value model. This involves losing the cyclic pressure pulsations and replacing the corresponding properties by cycle-averaged values. The multidimensional lookups that are needed for the mean value model are realized by a set of neural networks. These neural networks are trained with results of a design of experiments variation of the basic model over the complete operation range of the engine. The main flow parameters of the air path model as well as the engine speed are used as inputs to the neural networks. The outputs of the neural networks again serve as inputs into the flow model. Figure 3.13 shows the main structure of the information flow in the model. Depending on the model properties, the inputs into the neural networks can be extended to all major influences (gas composition, spark timing, valve timing).

The use of neural networks allows to reproduce the behavior of the detailed cylinder model to a very good extent. In the way they are used in this case, they replace multidimensional maps and offer some extended inter- and extrapolation functionalities. To ensure a good match between input and output data, the results from the detailed simulations are seperated into a training and a testing data set that are used consecutively.

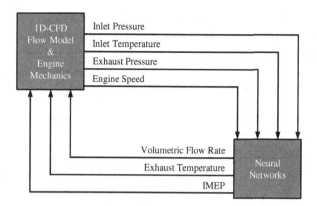

**Figure 3.13:** Information Flow in Mean Value Models

The regression plots in figure 3.14 show the almost perfect match of the neural networks outputs over the whole operation range. For the exhaust temperature, a clear separation can be seen between the fired and unfired operation points.

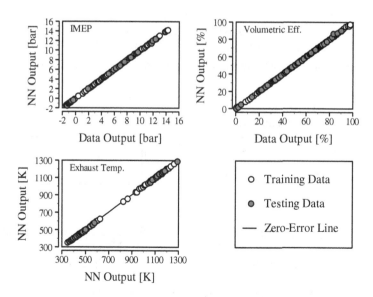

**Figure 3.14:** Regression Plots of Neural Network Training and Testing

A comparison of the pressure profiles with the FSM shows the effects of the different model approaches (figure 3.15). While the pressure profile of the FSM show the realistic pulsating behavior, the MVM is characterized by a continuous operation and averaged values. It is obvious that most of the dynamics of the airpath are lost when making this step of simplification.

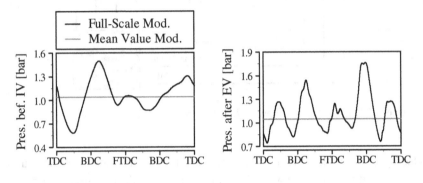

**Figure 3.15:** MVM Pressure Profiles Comparison (5000 rpm WOT)

A comparison of the resulting WOT torque curve (figure 3.16) with the basic model shows an error of still less than 3 % over the complete speed range and below 1.5 % for the most important operation range below 3000 rpm.

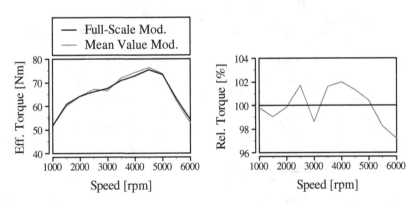

**Figure 3.16:** MVM Model Comparison of Torque at WOT

With the mean value cylinder model and the simplified air path of the FRM, another decrease in simulation time of more than 50 % compared to the FRM could be shown. With a real-time factor of 2, the mean value model (MVM) runs significantly faster even than the FRM.

### 3.2.6 Map-Based Models

Map-based engine models represent the most common way of ICE modeling in powertrain simulations. Due to the very quick computation times, the simulation speed of the complete vehicle model remains almost untouched. However, in contrast to the other models, the dynamic behavior of the ICE is completely neglected for this model. The only dynamic process that is taken into account is the inertia of the engine crank train. The dynamics of the air and fuel path and variations in cylinder load during transient operation are completely neglected.

The mapped ICE used in this work is described by a maximum load curve and maps for FMEP as well as fuel mass flow.

## 3.3 Engine Simulations

### 3.3.1 Stationary Comparison

Each of the described models was used to generate a consumption map to compare the stationary results before the models were used in driving cycle simulations. Figure 3.17 shows a comparison of the BSFC maps for FRM and FSM. Most of the operation range is reproduced at a very good level of accuracy with the deviations being below 0.5 %.

Only at operation points with very high or very low mass flows the deviations are higher than 0.5 %, but still below 3 %. The biggest differences are caused by a higher knock tendency due to small variations in cylinder filling.

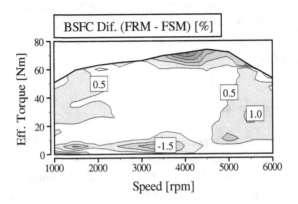

**Figure 3.17:** Relative Difference in Fuel Consumption: FRM vs. FSM

A comparison of the consumption values of the MVM with the FSM results shows a similar behavior with the biggest deviations also at very high and very low mass flows. However, in figure 3.18, a general tendency to higher errors compared to the FRM can be observed. The maximum error, at less than 5 %, is also higher than for the FRM but also limited to a very small range of the operation map.

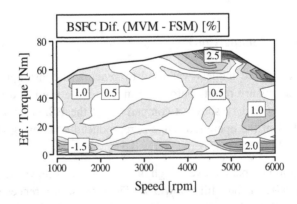

**Figure 3.18:** Relative Difference in Fuel Consumption: MVM vs. FSM

Generally, the consumption behavior of the engine is still reproduced with reasonable accuracy over the complete operation range. Especially the typical HEV operation area at low to medium speeds and medium to high torques is reproduced with very limited deviations of less than 1 %. It is obvious that there are no deviations for the map-based engine model in steady-state operation as the map is loaded with the exact same data.

The computation times of the stand alone models are characterized by real-time factors 100 for the full scale model, 5 for the fast running model and 2 for the mean value model. This shows that substantial gains in simulation speed can be realized with the presented approaches.

### 3.3.2 Dynamic Investigations

For a comparison of the dynamic responses of the different models, an engine startup with loadstep to nearly maximum load was investigated. This situation occurs on a frequent basis in a parallel hybrid powertrain. For the selected load step, the engine was accelerated to 2,000 rpm in 0.2 seconds. The requested torque after engine startup was set to 60 Nm. For all engine models the injection was enabled after exceeding 1,000 rpm which is referred to as hybrid start and can help to realize lower raw emissions and smoother engine start.

The simulations of FSM and FRM lead to virtually identical results, while MVM and MBM are very similar. The main difference between the model types in this closeup view is the cyclic operation of the FSM and FRM on the one hand side compared to the continuous operation of both MVM and MBM. An exemplary comparison of FSM and MVM is shown in figure 3.19.

In the first period of engine speed up, the responses of the models are almost identical, with only minor differences due the compression of the cylinders. The dynamics of the engine inertia are implemented identical in all of the four models. As soon as 1,000 rpm are reached and the injection is enabled, the MVM starts to build up torque, while the FSM needs another crankshaft rotation until the injected fuel reaches the cylinders. Due to this lag in engine torque, more mechanical energy is required to accelerate the engine. In the parallel hybrid vehicle setup, this energy must be provided by the electric machine and battery (generally at low efficiency due to the high required power), leading to an increase energy demand from the battery. Therefore, MVM and MBM tend to underestimate battery stress and fuel consumption.

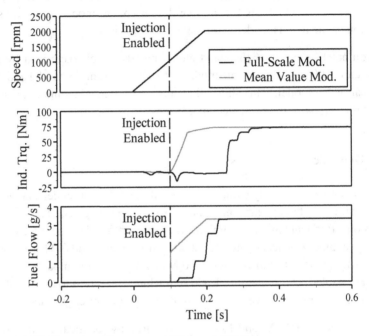

**Figure 3.19:** Transient Response: FSM vs. MVM

For a better comparison, table 3.4 shows some results of the simulations condensed to some simple parameters illustrating the differences.

**Table 3.4:** Load Step Simulation Results

|       | Time to 50 Nm | Req. Mech. Energy | Mean Eff. 2 Sec. |
|-------|---------------|-------------------|------------------|
| MBM   | 0.22 sec.     | 1152 J            | 30.6 %           |
| MVM   | 0.23 sec.     | 1220 J            | 30.6 %           |
| FRM   | 0.29 sec.     | 2504 J            | 28.7 %           |
| FSM   | 0.29 sec.     | 2505 J            | 28.9 %           |

The first column shows the time until the engine provides more than 50 Nm, where MBM and MVM show a startup time of only approximately 75 % compared to the other two models. Column 2 shows the mechanical energy required until the engine can provide positive torque. It can be found that the values for FSM and FRM are more than 100 % higher, compared to MBM and MVM. The third column presents the mean engine efficiency after the first 2 seconds of engine operation, for which the mechanical power is integrated (power required to accelerate the engine is here used as negative power) and divided by the integrated energy brought into the system from the injected fuel. Here, MBM and MVM show a higher mean efficiency than FRM and FSM due to the lower required mechanical energy for startup.

After this comparison of the engine models in standalone simulations, the next section provides the results of the vehicle simulations. While the boundary conditions for the engine models are significantly more dynamic than for the loadsteps in this section, the main drivers for the differences between the models are the same as described here.

# 3.4 Vehicle Simulations

## 3.4.1 Driving Cycles

When evaluating the SoC profiles and $CO_2$ emissions (or fuel consumption) for the vehicle simulations, the MBM and the MVM again show very similar behavior, which is why the FSM results are only compared to the map-based simulation results here. However, the most important results of the mean value model are included in the model comparison section below. Figure 3.20 shows the profiles of battery state of charge and cumulated $CO_2$ emissions.

Due to the dynamic responses of the FSM, more electric energy is needed during the urban part. In the extra urban part this has to be compensated by an increased load point shift which leads to higher $CO_2$ emissions at cycle end. This effect is very simmilar to what could be shown in the comparison of kinematic and dynamic drivetrain models in chapter 2. Again, it can be seen that even for the synthetic NEDC with its very low overall dynamic requirements, the results differ mainly in the urban part, where a frequent change between electric and hybrid operation is observed. The continuous drift for both total $CO_2$ and battery SoC can directly be related to the increased load point shift which is required to allow a balanced SoC.

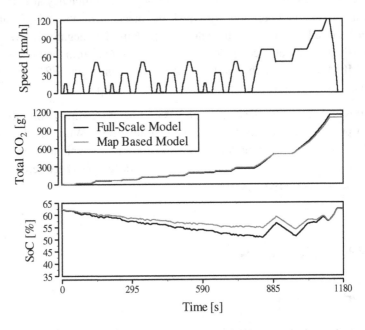

**Figure 3.20:** Battery SoC and $CO_2$ Emissions in NEDC

The higher dynamics of the AUDC lead to substantially increased transient requirements for the simulation. Figure 3.21 presents a comparison of the dynamic responses of the ICE models in the AUDC. It can be seen that even with the very short response time of the naturally aspirated engine, the FSM responds with short moments of lag, which have to be covered by an increased load on the electric machine, leading to a lower SoC as well as a higher fuel consumption for this part of the cycle. Interestingly, in all engine model comparisons, the main differences are always observed on the electrical path. In the extract from the AUDC here, the profiles for the fuel consumption are matching almost perfectly even though the engine models show a significantly different behavior. Nevertheless, the very fast reaction times of the electric propulsion system lead to the differences observed here.

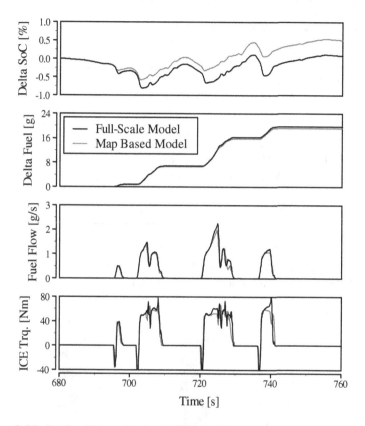

**Figure 3.21:** Engine Dynamics in AUDC

It has to be noted, that the results of simulations with such detailed models depend heavily on the quality of the implemented controllers for all actuators (throttle, injection, waste-gate etc.). While there was a lot of effort put into the optimization of these controllers, it is obvious that the quality of the ones in series production vehicles can still not be reached for all of these. This might have an influence on the results but could not be investigated here.

### 3.4.2 Model Comparison

The simulation results for the $CO_2$ emissions as well as the real-time factors for the described models are summarized in table 3.5.

**Table 3.5:** Driving Cycle Simulation Results for SCC Concept

| Cycle | Model | $CO_2$ emissions | Real-Time Factor |
|-------|-------|------------------|------------------|
| NEDC | MBM | 103.4 g/km (95.7 %) | 0.38 |
| | MVM | 104.5 g/km (96.7 %) | 4.6 |
| | FRM | 108.0 g/km (100 %) | 9.9 |
| | FSM | 108.0 g/km (100 %) | 231 |
| AUDC | MBM | 132.7 g/km (88.1 %) | 0.39 |
| | MVM | 135.4 g/km (89.9 %) | 4.2 |
| | FRM | 150.6 g/km (100 %) | 11 |
| | FSM | 150.4 g/km (99.9 %) | 212 |

An increase in computation time of almost 100 % compared to the real-time factors of the standalone engine models can be observed for all model types within the drivetrain simulations. A graphical overview of the results is shown in figure 3.22.

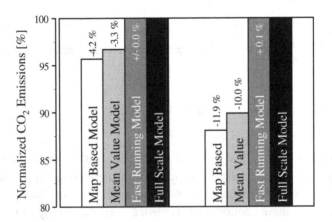

**Figure 3.22:** Model Comparison in Driving Cycle Simulations

A comparison of the $CO_2$ emissions shows the previously mentioned fact that FSM and FRM lead to virtually the same results, while MVM and MBM again deliver similar results. The simulation with FSM and FRM, however, leads to significantly higher emissions. For the NEDC, the differences of FSM versus MVM and FSM versus MBM are 3.3 % and 4.2 %, respectively. For the more dynamic AUDC, the differences are 10.0 % and 11.9 %, respectively.

While it is interesting that the shown differences can already be observed for a naturally aspirated engine, the bigger effects derive from the dynamics of turbo charged engines. Table 3.6 lists the results of the simulations for the MSV concept and the 15TC engine.

**Table 3.6:** Driving Cycle Simulation Results for MSV Concept with 15TC

| Cycle | Model | $CO_2$ emissions | Real-Time Factor |
|-------|-------|------------------|------------------|
| NEDC  | MBM   | 131.9 g/km (92.7 %) | 0.55 |
|       | MVM   | 134.5 g/km (94.6 %) | 3.6 |
|       | FRM   | 142.2 g/km (100 %) | 7.4 |
|       | FSM   | 142.1 g/km (100 %) | 97.8 |
| WLTC  | MBM   | 131.9 g/km (85.1 %) | 0.56 |
|       | MVM   | 134.3 g/km (86.6 %) | 3.2 |
|       | FRM   | 154.8 g/km (99.8 %) | 7.9 |
|       | FSM   | 155.0 g/km (100 %) | 94.4 |

It is obvious that the model differences are even bigger compared to the evaluations for the 08NA engine. Here, the distinct dynamics of the airpath play a very important role that can not be neglected. Also, when comparing the development of the real-time factors with table 3.5, it can be seen that the benefits of the faster models are not as big as before. This shows that the optimization potential for the computation time is highly dependent on the investigated engine concept. As the time step of the whole model may be restricted by only a single pipe, geometrical details are of high importance for the development of FRM or MVM. In some cases, basic 1D-CFD models already reach decent computation speeds, so that the optimization potential is farily low compared to other concepts. The comparison of the 08NA and 15TC engines above gives a good idea of these relationships.

## 3.5 Interpretation of the Results

A comparison of four different ICE models was presented. It was shown that all model types can produce a very good match in a stationary comparison. Transient comparisons showed that the continuously operating map-based and mean value models lead to different results than the cycle-based full scale and fast running models. While the more detailed full scale and fast running models provide a realistic reproduction of the transient response, the map-based and mean value models disregard important parts of the dynamic effects.

Within HEV drivetrain simulations it could be shown that in all simulations, the fast running model delivers virtually identical results as the full scale model, while taking only 5 - 8 % of the computation time. This indicates that fast running engine models can be very useful during vehicle development. On the other hand, the mean value model delivers almost the same results as the map-based model. Compared to the fast running and full scale model, these models calculated 3 - 15 % lower $CO_2$ emissions, depending on the simulated driving cycle.

The computation times of the vehicle models resulted in real-time factors of 0.38 - 0.56 with the map-based model, 3.2 - 4.6 with the mean value model, 7.4 – 10.9 with the fast running model and 94.4 – 231 with the full scale model.

The results indicate that, even with a naturally aspirated engine, the use of a 'traditional' map-based engine model within a dynamic HEV drivetrain can lead to a significant loss in simulation quality compared to a detailed 1D-CFD model. Also, if a well tuned fast running model is available, it can reduce the required simulation time significantly compared to a full scale model. As the computation time of a fast running model is still substantially higher than for a map-based approach (approximately factor 15-25), the results suggest using a simple, map-based model for high-level operation strategy development, but validating the results by simulations with a well tuned 1D-CFD model, preferably a fast running model. The fast running model should also be the tool of choice when more accurate results or a more detailed examination of the dynamic effects are required.

# 4 Battery

As the secondary energy storage, batteries are key components in hybrid power-trains. Even for charge sustaining hybrids, where ultimately all energy comes from fuel, the efficiency and the dynamic behavior of the electric energy storage has a significant impact on the fuel consumption and driving performance. However, the availability of detailed models is often problematic which is why it is important to understand the impact of battery models with different complexity on powertrain simulations.

This chapter presents an analysis of transient effects in battery models and their impact on the simulation of hybrid electric powertrains. First, the fundamentals of batteries and battery modeling are presented, before the simulation results are shown.

## 4.1 Batteries for Automotive Applications

Batteries were used in passenger cars since the very beginnings of the auto-motive industry. While the use of batteries in conventional vehicles is mostly limited to providing the energy to start the combustion engine, the require-ments are significantly higher for hybrid powertrains. Therefore, big improve-ments in battery technology were made in the last decade. There are basically three battery types that are used in automotive applications today which are Lead Acid (PbA), Nickel-Metal Hydride (NiMH) and Lithium Ion (Li-Ion). Table 4.1 gives an overview of the main advantages an disadvantages. [47]

Lead acid technology is well established for starter batteries in conventional vehicles. Lead acid batteries can be produced at relatively low cost and fulfill the highest safety standards. However, they perform very poorly at low temper-atures and provide only low gravimetric and volumetric energy density. Nickel-Metal Hydride batteries offer higher specific energy and power compared to lead acid. Also, they fulfill high safety standards since they are very tolerant

© Springer Fachmedien Wiesbaden GmbH, part of Springer Nature 2019
F. Winke, *Transient Effects in Simulations of Hybrid Electric Drivetrains*, Wissenschaftliche Reihe Fahrzeugtechnik Universität Stuttgart, https://doi.org/10.1007/978-3-658-22554-4_4

to abuse, the disadvantages are relatively high cost and low efficiency. NiMH technology was used in the first modern production HEV that made significant sales, for example the first versions of the Toyota Prius. [41] Since then, the development of Lithium Ion batteries made big improvements, especially driven by the demand for higher energy density. Li-Ion batteries provide one of the highest energy densities, high efficiency, no memory effect and only very low self-discharge and have thus become the most common battery technology in consumer electronics. The main disadvantages are the high cost and safety issues. When overheated, Li-Ion batteries can suffer from a phenomenon called thermal runaway, which can lead to batteries exploding or catching fire. However, Li-Ion batteries are also gaining importance in automotive applications and are considered to have high potential for future optimization. [47]

**Table 4.1:** Overview of Main Automotive Battery Types [13, 47]

| Battery Type | Advantages | Disadvantages |
|---|---|---|
| Lead Acid (PbA) | Low Cost<br>Safe<br>Established | Cold Temp. Performance<br>Short Life<br>Specific Energy |
| Nickel-Metal Hydride (NiMH) | Specific Energy (+)<br>Specific Power (+)<br>Abuse-Tolerant | High Cost<br>Low Efficiency<br>High Temp. Performance |
| Lithium Ion (Li-Ion) | Specific Energy (++)<br>Specific Power (++)<br>Low Self-Discharge | Safety<br>High Cost<br>Short Life |

Future developments in battery technology are expected to further increase the performance especially in terms of specific energy, cost and life span. Also higher degrees of battery safety can be expected. Promising developments include improved battery chemistry which may help to increase the specific energy of Li-Ion technology on the battery pack level (including casing and cooling) from approximately 180 Wh/kg to 240 Wh/kg or more. Other chemistries like Metal-Air Cells have even higher theoretical potential but have practical limitations that do not allow the full usage of their potential. [13]

While recent developments in battery technology allowed big improvements in all sectors, purely electrical propulsion systems still suffer from high battery cost, problems with the calendar and cycle life and long charging times. Therefore, it can be expected that internal combustion engines will remain at a significant market share in the medium term. Meanwhile, hybrid electric vehicles offer a solution that combines some of the advantages of both concepts.

## 4.2 Battery Modeling

Several approaches can be used for the modeling of the static and dynamic behavior of batteries. The most common classification of electrical models separates two groups, *first-principle models* and *empirical models*. [13]

First-principle models attempt to predict the battery behavior from a detailed description of the fundamental processes, like electrochemistry, thermodynamics and mass transport. This is often described as bottom-up approach. Such models are very detailed and specific for a given cell type, but need a lot of information on the material properties and cell structure. Although this model type is mainly based on generally available data (thermodynamic and chemical properties), they are usually complex to set up and calibrate. [13]

Empirical models follow a top-down approach and characterize the input and output behavior based on experimental data, regardless of the physics and chemistry involved. They are not predictive, but intuitive and simple to build and use. These models are well-suited when the general behavior of the battery system shall be represented, but no detailed understanding of the underlying processes is required. The best known approach for predicting battery behavior are *Equivalent Circuit Models*, which were used in this work. [13]

The basic idea of equivalent circuit models is to replicate the behavior of real components by a combination of ideal elements such as voltage sources, resistances or capacitances. All necessary parameters are then taken from map and table data as a function of the most important boundary conditions, which generally are temperature and state of charge. In some cases it may be necessary to use separate maps for charging and discharging. For the modeling of a battery cell, the simplest case can be realized by an ideal voltage source that provides the open circuit voltage $V_{OC}$ and an internal resistance $R_0$ that approximates the losses within the battery cell as shown in 4.1. This approach does not include cell dynamics, but it can therefore be calibrated relatively simple with only static measurements. [26]

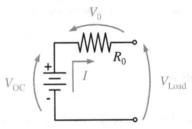

**Figure 4.1:** Static Battery Modeling

Using Kirchoff's voltage law, the cell terminal voltage $V_{Load}$ can be calculated as a function of the battery current as

$$V_{Load}(t) = V_{OC} - R_0 \cdot I(t).$$                    eq. 4.1

The state of charge can be determined from integration of the cell current as

$$SoC(t) = SoC_0 - \int_{t_0}^{t} \frac{I(t)}{C_{nom}} \, dt ,$$                    eq. 4.2

where $SoC_0$ is the initial state of charge and $C_{nom}$ the nominal capacity which is also a function of temperature. The sign of the current is defined such that $I$ is positive during discharging and negative during charging.

Note that the calculation of the SoC based on integration is only valid if the current and nominal capacity are known parameters at a high level of accuracy. For simulation models this is the case, but in reality, where measurement error and noise affect this calculation, state of charge estimation is more complicated and is generally an important part of the battery management. [43]

If dynamic effects shall be included, the model shown in figure 4.1 can be extended with a number of R-C branches in series. This allows to account for dynamic responses with different timescales, the time constant for each R-C branch is $\tau = R_n \cdot C_n$. A model with $n$ R-C branches is called $n^{\text{th}}$ order, so that the static model in figure 4.1 is a $0^{\text{th}}$ order model, while a model with two R-C branches would be $2^{\text{nd}}$ order. Figure 4.2 illustrates the equivalent circuit diagram of an $n^{\text{th}}$ order model. [13, 26]

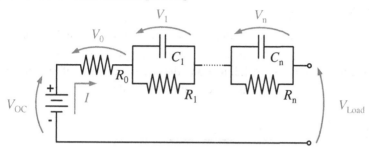

**Figure 4.2:** Dynamic Battery Modeling

The better accuracy of the model comes at the price of significantly higher calibration effort. The values of the parameters are most commonly estimated by curve fitting of experimental data. As they are highly variable with the operating conditions, the fitting becomes much more complex with increasing order, while the number of parameters also increases. For a model of $n^{\text{th}}$ order, the terminal voltage can be calculated as

$$V_{Load}(t) = V_{OC} - R_0 \cdot I(t) - \sum_{i=1}^{n} V_i, \qquad \text{eq. 4.3}$$

with the voltage at each R-C branch is calculated as

$$V_i(t) = V_{i,0} + \int_{t_0}^{t} \frac{1}{C_i} \left( I - \frac{V_i}{R_i} \right) dt. \qquad \text{eq. 4.4}$$

All values are evaluated on cell level, while battery packs in HEV consist of a number of cells connected in series (strings) and usually several strings in parallel. The values for the whole battery pack can be calculated from the cell values as

$$V_{Batt}(t) = V_{Cell}(t) \cdot n_{ser}, \qquad \text{eq. 4.5}$$

$$I_{Batt}(t) = I_{Cell}(t) \cdot n_{par}, \qquad \text{eq. 4.6}$$

where $n_{ser}$ is the number of cells in each string and $n_{par}$ the number of strings in parallel. This shows that a desired nominal voltage can be realized by selecting the corresponding number of cells in series, while the maximal available currents can be influenced by increasing or decreasing the number of strings in parallel.

In addition to a correct modeling of the electrical circuit, it is important to simulate the thermal behavior of the battery system. The model used in this work is illustrated in figure 4.3, it shows the two major flows of thermal energy which are the battery losses and the heat transferred to the cooler.

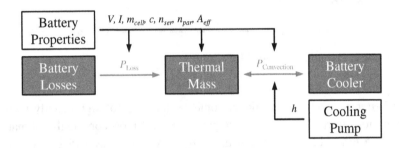

**Figure 4.3:** Battery Thermal Modeling

The battery losses can be calculated from the electrical model as

$$P_{Loss}(t) = I(t) \cdot \big( V_{OC}(t) - V_{Load}(t) \big), \qquad \text{eq. 4.7}$$

where all values are for the whole battery pack and can be calculated from cell level using eq. 4.5 and eq. 4.6.

The convective heat flow to the cooling medium is calculated as

$$P_{Convection}(t) = h \cdot A_{eff} \big( T_{Batt}(t) - T_{Amb} \big), \qquad \text{eq. 4.8}$$

where $h$ is the convective heat transfer coefficient, $A_{eff}$ is the effective area exposed to the cooling medium and $T_{Batt}$ and $T_{Amb}$ are the temperatures of the battery and ambient, respectively. The heat transfer coefficient $h$ is controlled by a cooling pump, so that it can be increased for higher battery temperatures to increase cooling. $A_{eff}$ depends on the cell and battery structure and the number of cells that are used. Assuming a perfect cooler, $T_{Amb}$ is used instead of the actual temperature of the cooling medium, so that the cooling circuit does not have to be calculated separately.

Using the first principle, the temperature of the battery can be calculated from the incoming and outgoing power flows. With the signs of the power flows such that incoming power flows are positive, the temperature becomes

$$T_{Batt}(t) = T_{Batt,0} + \frac{1}{m_{Batt} \cdot c} \cdot \int_{t_0}^{t} \left( P_{Loss}(t) - P_{Convection}(t) \right) dt, \qquad \text{eq. 4.9}$$

where $T_{Batt,0}$ is the initial temperature of the battery and $m_{Batt}$ and $c$ are the mass and heat capacity of the battery, respectively.

## 4.3 Used Battery Models

The battery data used for the simulations in this work is based on measurements of Li-Ion cells optimized for automotive applications. Some of the main parameters of the cell are listed in table 4.2.

**Table 4.2:** Main Parameters of the Used Li-Ion Cell

| | |
|---|---|
| Cell Weight | 70 g |
| Nominal Voltage | 3.3 V |
| Nominal Capacity | 2.5 Ah |
| Max. Continuous Current | 70 A |
| Max. Peak Current | 120 A |

To realize the operating voltage of the electric machine, 81 cells are combined to a string for all setups. The number of parallel strings in the battery pack depends on the investigated system. For the basic 21 kW electric machine as shown in the subcompact concept in the previous chapter, three parallel strings are used for a nominal capacity of 7.5 Ah and an operating voltage of 267 V. For the 27 kW electric machine of the other concepts, a fourth string is added to the battery pack.

### 4.3.1 Dynamic Model

Due to the available cell data, a first order dynamic model was used in this work. While this model type is capable to predict the main dynamic effects of the battery behavior, the required parameters are limited to the open circuit voltage $V_{OC}$, two resistances $R_0$ and $R_1$ as well as one capacitance $C_1$. The open circuit voltage is a function of the state of charge, all other parameters are functions of the state of charge, battery temperature and the current direction. Figure 4.4 shows the corresponding curve and maps for discharging.

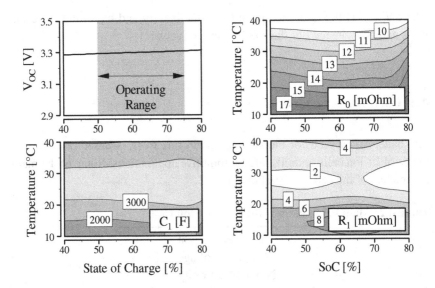

**Figure 4.4:** Dynamic Cell Model Parameters

It can be seen that $V_{OC}$ shows the typical flat gradient in the operating range, while higher gradients only occur towards the end points of the SoC. The operating range is not only restricted to limit the variance in terminal voltage, but also to restrict battery aging which is an important issue in HEV, due to the aggressive loading cycles. As an operation of the battery at high temperatures and close to the end points of the SoC accelerate battery aging, the available SoC range is typically limited for HEV as shown in figure 4.4.

### 4.3.2 Derivation of a Static Model

To ensure maximum comparability, a static model was derived from the same data as the dynamic model. For the static case, the internal resistance of the dynamic battery model ultimately comes to

$$R_i = R_0 + R_1,$$ 
eq. 4.10

so that the value of the internal resistance for the static model can be determined directly from the parameters of the dynamic model. The calculation of the open circuit voltage is not affected, the parameters are shown in figure 4.5.

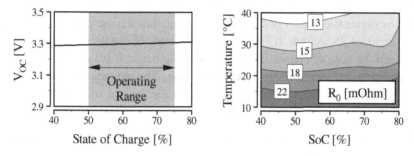

**Figure 4.5:** Static Cell Model Parameters

For static voltage modeling, the circuit can be calculated from a closed equation as

$$P_{Load} = V_{Load} \cdot I = V_{OC} \cdot I - R_0 \cdot I^2.$$ 
eq. 4.11

The current can then be calculated directly as a function of power as

$$I = \frac{V_{OC} - \sqrt{V_{OC}^2 - 4 \cdot R_0 \cdot P_{Load}}}{2 \cdot R_0}.$$ 
eq. 4.12

This approach is also used to apply current limitations in the operation strategy, even if the actual battery model is dynamic.

## 4.4 Battery Simulations

Before the battery models were used and investigated in the powertrain environment, some standalone simulations were evaluated. For this purpose, the thermal and electrical model were separated so that the dynamic influence of each part of the model could be identified.

The transient effects of the electrical and thermal model make their impact on different timescales. While the dynamics of the electrical model influence the behavior within several seconds before reaching steady-state, the temperature dependence of the battery properties becomes important on a timescale of several minutes or, for example, over a complete driving cycle.

To investigate the impact of the electrical dynamics, a pulse profile is applied that includes 25 seconds at 10 kW and 20 seconds of rest between each pulse. To keep the state of charge within a reasonable range, it is switched between discharging and charging mode for each pulse. The temperature was set to a constant 20°C to avoid cross-influences from the thermal model. Figure 4.6 illustrates the results of the simulations.

The dynamics of the electrical model lead to differences in terminal voltage and losses. While the static model directly comes to steady-state, the voltage and losses of the dynamic model lag behind leading to lower losses compared to the static approach. This effect can be chemically explained by a change in ion-concentration inside the electrolyte near the electrodes which leads to an increase of the voltage drop and thus to higher losses.

Interestingly, in this case, the static model overestimates the losses leading to a drift of SoC which is in contrast to the results for all dynamic engine models in the chapter before. This can lead to a compensation of errors, if a vehicle simulation with static battery and combustion engine is compared to one with both components modeled dynamically. When comparing different simulation setups, this has to be taken into account. For the investigations in this work, a static (map based) engine model was used for all battery related simulations in order to ensure a minimum impact of other components on the results.

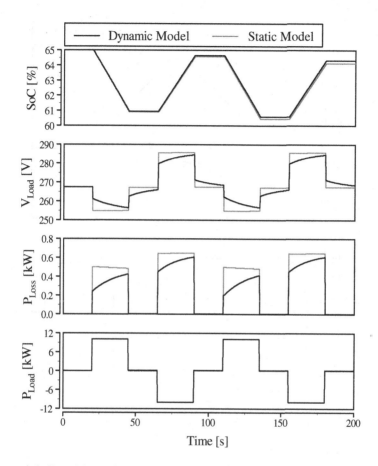

**Figure 4.6:** Standalone Comparison of Dynamic and Static Battery Model

In a second comparison the dynamic model was used with thermal model and compared to the identical electric model with fixed temperature. Figure 4.7 illustrates the results.

While the increase of battery temperature is only moderate in the evaluated section, a significant change of the losses can be observed. The raised temperatures lead to increased activity of the electrolyte and faster chemical reactions, resulting in decreasing losses in the battery cell.

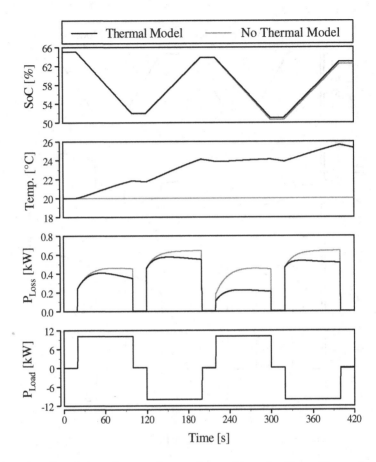

**Figure 4.7:** Standalone Battery Comparison with and without Thermal
          Model

The following section presents an investigation of the different models within
the environment of the parallel HEV powertrain. The evaluations are presented
for the urban hybrid concept which has the highest hybridization rate and thus
the biggest impact of the electric system.

## 4.5 Vehicle Simulations

When comparing the results of the vehicle simulations, it can be seen that the differences of the battery models are small enough that the decisions of the operation strategy are affected to only a small extend, at least for the NEDC. The upper part of figure 4.8 shows that the terminal power of the battery is almost unaffected. The models that were used here were the dynamic battery model with thermal modeling and the static model with constant ambient temperature. While the terminal power is almost equivalent, the losses show the same differences that can be observed with the standalone comparisons above.

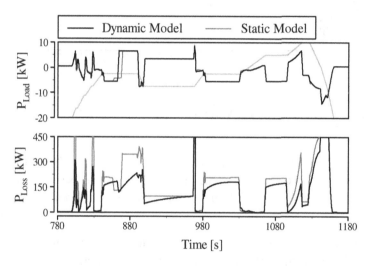

**Figure 4.8:** Battery Terminal Power and Losses during NEDC

The trailing behavior is due to the electrical dynamics and the different levels when close to steady-state due to the thermal behavior. The temperature profiles for NEDC and AUDC are shown in figure 4.9.

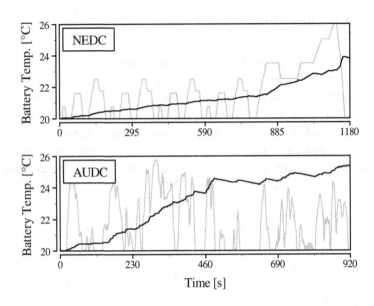

**Figure 4.9:** Battery Terminal Power and Losses during NEDC

The resulting temperature rises are relatively moderate, with $\Delta T \approx 4\ C$ for the NEDC and $\Delta T \approx 5\ C$ for the AUDC. Nevertheless, an increase of the cell temperature from 20 °C to 25 °C leads to a decrease of the internal resistance of approximately 10 %.

A comparison of the integrated losses of the battery shows big differences. Figure 4.10 illustrates the battery share of the total powertrain losses, which are electric machine, transmission, final drive, clutches, rolling and aerodynamic resistances and auxiliary loads. The only exception is the internal combustion engine, which for this consideration is seen as power source. It is not surprising that the differences are relatively small for the NEDC with its quasi-static driving profile and the long phases of steady-state operation. Although there are already differences as shown above, they do not translate to significant impacts on the overall results. For the urban operation of the AUDC, however, the impact of the electric path is very important so that the battery losses play a considerable role in the overall powertrain losses.

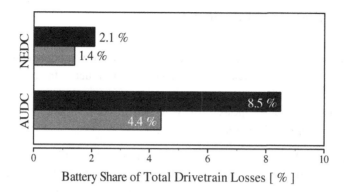

Battery Share of Total Drivetrain Losses [ % ]

**Figure 4.10:** Battery Losses for NEDC and AUDC

A comparison of the resulting $CO_2$ emissions is shown in figure 4.11. As expected from the analyses above, the differences in the results are significantly higher for the more dynamic profile of the AUDC. While the results are comparable for the synthetic NEDC, the effect that the dynamic battery model has on the overall results is of a considerable magnitude.

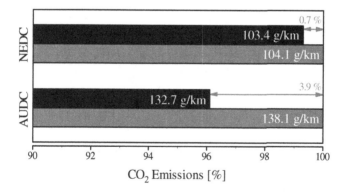

$CO_2$ Emissions [%]

**Figure 4.11:** $CO_2$ Emissions for NEDC and AUDC

## 4.6 Interpretation of the Results

A standalone comparison of dynamic and static battery models showed that the modeling of the electrical and thermal behavior influence the results for the battery losses as well as the state of charge to a significant extent. Therefore, simulations of the HEV powertrain with the different models were used to evaluate the impact on the integrated system.

The driving cycle simulation results show that the choice of battery model can have substantial impact on the quality of simulation results for parallel HEV. While the static battery model works reasonably well for the synthetic profile of the NEDC, as soon as a more realistic driving profile as the AUDC is chosen, the differences may not be omitted. The computation times of the models were not notably affected by the different model structures, as the computation time for the battery model can be neglected compared to the overall powertrain model.

Therefore it is recommended to work with a dynamic battery model coupled with a thermal model whenever the availability of data permits.

# 5 Summary and Conclusion

For this work, a range of component and powertrain models for HEV were developed. The main goal was to provide an overview of possible modeling approaches and their impact on the results of consumption and $CO_2$ oriented simulations. In order to understand the underlying mechanisms and to analyze varying boundary conditions, a set of engine and vehicle concepts was developed that stretches from a Sub-Compact Car to a fullsize SUV and from a naturally aspirated 2-cylinder MPI engine to a turbo charged 4-cylinder DI engine (all SI engines). Also, a detailed energetic analysis of the used driving cycles was presented so that the similarities and differences of the speed profiles can be evaluated.

The investigations on kinematic and dynamic powertrain models showed that the selection between these modeling approaches has a major impact on the calculated $CO_2$ emissions. While the results depend highly on the boundary conditions, the projections with the kinematic approach are up to 20 % lower than the dynamic model. These differences are due to the fact that the kinematic approach can not capture transient processes which for HEV are mostly during the engine startup after pure electric driving. As the additional modeling effort and the increases in simulation time are relatively low, it can be concluded that a dynamic powertrain model should generally be preferred over a kinematic model whenever the boundary conditions permit.

For the combustion engine, four different modeling approaches were analyzed. A conventional 1D-CFD model served as the basis for all following investigations. Fast running engine models were developed through optimizations of the airpath. In a further step of simplification, the model of the combustion chamber was replaced with a mean value based approach with the help of neural nets. A map-based ICE model represents the traditional method of engine modeling within powertrain simulations. It could be shown that all of the investigated model types are able to provide very good results in stationary comparisons. However, for transient operation and especially in driving cycle simulations the continuously operating mean value and map-based mod-

© Springer Fachmedien Wiesbaden GmbH, part of Springer Nature 2019
F. Winke, *Transient Effects in Simulations of Hybrid Electric Drivetrains*, Wissenschaftliche
Reihe Fahrzeugtechnik Universität Stuttgart, https://doi.org/10.1007/978-3-658-22554-4_5

els show deviations compared to the fast running and full scale model. The investigated fast running models delivered virtually identical results as the respective full scale models. While the possible reduction of computation time with fast running models was in the range of 95 %, this approach still offers the full functionalities of conventional models. Thus, it could be shown that fast running engine models provide a powerful tool for HEV powertrain simulations.

Investigations of the battery dynamics showed a significant influence on the overall results of HEV simulations. In standalone comparisons of dynamic and static battery models, it was shown that both the electrical and the thermal dynamics of the battery make an impact that should not be neglected. As the computation times of the models are not notably affected by the different model structures, the deciding criterion here is generally the availability of data. The conclusion that can be drawn from the analyses in this work is that a dynamic battery model coupled with a well tuned thermal model should be used whenever the required data is available.

For future works, it would make sense to analyze the impact of the low-level controls (combustion engine, clutches, brake distribution) on the results. While the concept for this work was to use the best possible controls (minimized reaction times), the control structure for the production components may differ in some ways from the structure of the models here. In some cases restrictions of the control units in the production powertrain may impact the results significantly. To realize the best possible simulation results it is necessary to have very detailed knowledge about the used control structure. Also, the used database could be improved, especially with more transient measurements on a dynamic engine test bench or on a chassis dynamometer. The most efficient development process can only be ensured when simulation and measurement are working closely together so that the full potential of both major development tools may be used.

# Bibliography

[1] M. André. Real-world driving cycles for measuring cars pollutant emissions - Part A: The ARTEMIS European driving cycles, 2004.

[2] M. André. Real-world driving cycles for measuring cars pollutant emissions - Part B: Driving cycles according to vehicle power, 2004.

[3] M. Auerbach. *Phlegmatisierung des Dieselmotors im Hybridverbund.* PhD thesis, Universität Stuttgart, 2012.

[4] M. Bargende. *Ein Gleichungsansatz zur Berechnung der instationären Wandwärmeübergänge im Hochdruckteil von Ottomotoren.* PhD thesis, Technische Hochschule Darmstadt, 1991.

[5] M. Bargende. Berechnung und Analyse innermotorischer Vorgänge. lecture notes, 2013.

[6] M. Bargende, U. Riegler, and B. Scholz. Der virtuelle Motor - Fiktion oder Realität? In *Tagung im Haus der Technik - Essen*, 2000.

[7] R. v. Basshuysen. *Natural Gas and Renewable Methane for Powertrains - Future Strategies for a Climate-Neutral Mobility.* Springer, Berlin, Heidelberg, 1st ed. 2016 edition, 2016.

[8] R. v. Basshuysen and F. Schäfer. *Handbuch Verbrennungsmotor - Grundlagen, Komponenten, Systeme, Perspektiven.* Springer-Verlag, Berlin Heidelberg New York, 7. aufl. edition, 2014.

[9] R. E. Bellman. *Dynamic Programming.* Princeton University Press, 1957.

[10] D. P. Bertsekas. *Dynamic Programming and Optimal Control.* Athena Scientific, 2nd edition, 2000.

[11] T. Böhm. Erfahrungen mit dem VW TwinDrive. In *5. E-Motive Expertenforum Elektrische Fahrzeugantriebe*, Stuttgart, 2012.

© Springer Fachmedien Wiesbaden GmbH, part of Springer Nature 2019
F. Winke, *Transient Effects in Simulations of Hybrid Electric Drivetrains*, Wissenschaftliche Reihe Fahrzeugtechnik Universität Stuttgart, https://doi.org/10.1007/978-3-658-22554-4

[12] D. Boland. *Wirkungsgradoptimaler Betrieb eines aufgeladenen 1,0 l Dreizylinder CNG Ottomotors innerhalb einer parallelen Hybridarchitektur.* PhD thesis, Universität Stuttgart, 2011.

[13] M. Canova. Battery and fuel cell systems for electrified vehicles. Stuttgart International Summer School Mobility, 2015.

[14] M. Chiodi. *An Innovative 3D-CFD-Approach towards Virtual Development of Internal Combustion Engines.* PhD thesis, Universität Stuttgart, 2011.

[15] Council of the European Union. Council Directive 70/220/EEC of 20 March 1970 on the approximation of the laws of the Member States relating to measures to be taken against air pollution by gases from positive-ignition engines of motor vehicles, 1970.

[16] R. Courant, K. Friedrichs, and H. Lewy. Über die partiellen differenzengleichungen der mathematischen physik. *Mathematische Annalen*, 100(1):32–74, 1928.

[17] DIN 1343. Referenzzustand, Normzustand, Normvolumen - Begriffe und Werte, 1990.

[18] F. Dudenhöffer. Die Marktentwicklung von Hybrid-Fahrzeugkonzepten. *Automobiltechnische Zeitschrift (ATZ)*, 04, 2005.

[19] M. Duoba, H. Ng, and R. Larsen. Characterization and Comparison of Two Hybrid Electric Vehicles (HEVs) - Honda Insight and Toyota Prius. In *SAE Technical Paper*. SAE International, 03 2001.

[20] Economic Commission for Europe of the United Nations (UN/ECE). Regulation No 101, 2010.

[21] European Environment Agency. Annual European Union greenhouse gas inventory 1990–2014 and inventory report, 2016.

[22] A. Froberg. Inverse dynamic simulation of non-quadratic mimo powertrain models - application to hybrid vehicles. In *2006 IEEE Vehicle Power and Propulsion Conference*, pages 1–6, 2006.

[23] C. Gray. Hydraulic Hybrid Vehicle. United States Patent US6719080, 2004.

[24] M. Grill. *Objektorientierte Prozessrechnung von Verbrennungsmotoren.* PhD thesis, Universität Stuttgart, 2006.

[25] D. Gross, W. Ehlers, P. Wriggers, J. Schröder, and R. Müller. *Dynamics – Formulas and Problems - Engineering Mechanics 3.* Springer, Berlin, Heidelberg, 2016.

[26] L. Guzzella and A. Sciarretta. *Vehicle Propulsion Systems - Introduction to Modeling and Optimization.* Springer Science & Business Media, Berlin Heidelberg, 2012.

[27] B. Hartmann and C. Renner. Autark, Plug-In oder Range Extender? Ein simulationsgestützter Vergleich aktueller Hybridfahrzeugkonzepte. In *18. Aachener Kolloquium Fahrzeug- und Motorentechnik*, 2009.

[28] B. Heißing, M. Ersoy, S. E. Gies, B. Heißing, M. Ersoy, and S. Gies. *Fahrwerkhandbuch - Grundlagen · Fahrdynamik · Komponenten · Systeme · Mechatronik · Perspektiven.* Springer-Verlag, Berlin Heidelberg New York, 4. aufl. edition, 2013.

[29] P. Hofmann. *Hybridfahrzeuge - Ein alternatives Antriebssystem für die Zukunft.* Springer-Verlag, Berlin Heidelberg New York, 2. aufl. edition, 2014.

[30] G. Hohenberg. *Experimentelle Erfassung der Wandwärme in Kolbenmotoren.* Habilitation thesis, Technische Universität Graz, 1980.

[31] F. P. Incropera and D. P. DeWitt. *Fundamentals of heat and mass transfer.* John Wiley & Sons Australia, Limited, Hoboken, 5. aufl. edition, 2002.

[32] P. d. Jägher. *Einfluss der Stoffeigenschaften der Verbrennungsgase auf die Motorprozessrechnung.* PhD thesis, Technische Universität Graz, 1984.

[33] Japanese Industrial Safety and Health Association. Jisha 899, 1983.

[34] E. Justi. *Spezifische Wärme Enthalpie, Entropie und Dissoziation technischer Gase.* Springer-Verlag, Berlin Heidelberg New York, 1938.

[35] T. Köppen. Die Rolle der Firma Jakob Lohner & Co. bei der Entwicklung von Hybridantrieben im Automobilbau. *Technikgeschichte*, Bd. 55, 1988.

[36] Kraftfahrt-Bundesamt. Fahrzeugbestand in den Jahren 1960 bis 2017 nach Fahrzeugklassen, 2017.

[37] K. Küpper, A. Weinzerl, and N. Dumont. Efficient Powertrain Solutions for 12 V up to 800 V. In *Shanghai-Stuttgart-Symposium Automotive and Engine Technology*, 2016.

[38] S. Lecheler. *Numerische Strömungsberechnung*. Vieweg+Teubner Verlag, 2011.

[39] G. P. Merker and R. Teichmann. *Grundlagen Verbrennungsmotoren - Funktionsweise, Simulation, Messtechnik.* Springer-Verlag, Berlin Heidelberg New York, 7. aufl. edition, 2014.

[40] U. Mückenberger and S. Timpf. *Zukünfte der europäischen Stadt: Ergebnisse einer Enquete zur Entwicklung und Gestaltung urbaner Zeiten.* VS Verlag für Sozialwissenschaften, 2007.

[41] D. Naunin. *Hybrid-, Batterie- und Brennstoffzellen-Elektrofahrzeuge - Technik, Strukturen und Entwicklungen.* expert verlag, Renningen, 4. aufl. edition, 2007.

[42] B. Noll. *Numerische Strömungsmechanik*. Springer-Verlag Berlin Heidelberg New York, 1993.

[43] S. Onori, L. Serrao, and G. Rizzoni. *Hybrid Electric Vehicles - Energy Management Strategies.* Springer, Berlin, Heidelberg, 1st ed. 2016 edition, 2015.

[44] G. Paganelli, S. Delprat, T. M. Guerra, J. Rimaux, and J.-J. Santin. Equivalent consumption minimization strategy for parallel hybrid powertrains. In *IEEE 55th Vehicular Technology Conference*, volume 4, 2002.

[45] H. Pieper. Mixed Drive for Autovehicles. United States Patent US0913846, 1909.

[46] R. Pischinger, M. Klell, and T. Sams. *Thermodynamik der Verbrennung-skraftmaschine.* Springer-Verlag, Berlin Heidelberg New York, 3. aufl. edition, 2013.

[47] C. D. Rahn and C.-Y. Wang. *Battery Systems Engineering.* John Wiley & Sons Ltd, 2013.

[48] K. Reif. *Konventioneller Antriebsstrang und Hybridantriebe - mit Brennstoffzellen und alternativen Kraftstoffen.* Springer-Verlag, Berlin Heidelberg New York, 1. aufl. edition, 2010.

[49] K. Reif, K. E. Noreikat, and K. Borgeest. *Kraftfahrzeug-Hybridantriebe - Grundlagen, Komponenten, Systeme, Anwendungen.* Springer-Verlag, Berlin Heidelberg New York, 2012. aufl. edition, 2012.

[50] C. Reulein. *Simulation des instationären Warmlaufverhaltens von Verbrennungsmotoren.* PhD thesis, Technische Universität München, 1998.

[51] T. Riemer. *Vorausschauende Betriebsstrategie für ein Erdgashybridfahrzeug.* PhD thesis, Universität Stuttgart, 2012.

[52] M. Ruf. *Potentiale des Dieselhybrids durch optimierte Betriebsstrategie.* PhD thesis, Universität Stuttgart, 2012.

[53] F. R. Salmasi. Control strategies for hybrid electric vehicles: Evolution, classification, comparison, and future trends. *IEEE Transactions on Vehicular Technology*, 56(5):2393–2404, Sept 2007.

[54] H. Schmidt. Worldwide Harmonized Light-Vehicles Test Procedure (WLTP) und Real Driving Emissions (RDE) – aktueller Stand der Diskussion und erste Messergebnisse. In M. Bargende, H.-C. Reuss, and J. Wiedemann, editors, *15. Internationales Stuttgarter Symposium: Automobil- und Motorentechnik*, 2015.

[55] M. Schwarzmeier. *Der Einfluss des Arbeitsprozessverlaufs auf den Reibmitteldruck von Dieselmotoren.* PhD thesis, Technische Universität München, 1992.

[56] A. Sciaretta, L. Guzzella, and C. H. Onder. On the power split control of parallel hybrid vehicles: from global optimization towards real-time control. *Automatisierungstechnik*, 51(5):195–203, 2003.

[57] A. Sciarretta, M. Back, and L. Guzzella. Optimal control of parallel hybrid electric vehicles. *IEEE Transactions on control systems technology*, 12(3), 2004.

[58] A. Sciarretta and L. Guzzella. Control of hybrid electric vehicles. *IEEE Control Systems*, 27(2):60–70, April 2007.

[59] L. Serrao. *A comparative analysis of energy management strategies for hybrid electric vehicles*. PhD thesis, The Ohio State University, 2009.

[60] C. Soanes and A. Stevenson. *Concise Oxford English Dictionary - Luxury Edition*. OUP Oxford, New York, London, 12 con ind edition, 2011.

[61] S. Stan. *Alternative Antriebe für Automobile*. Springer-Verlag, Berlin Heidelberg New York, 4. aufl. edition, 2015.

[62] M. Stiegeler. *Entwurf einer vorausschauenden Betriebsstrategie für parallele hybride Antriebsstränge*. PhD thesis, Universität Ulm, 2008.

[63] Umweltbundesamt. Annual Greenhouse Gas Emissions in Germany by Category, 2016.

[64] UN. Kyoto protocol to the united nations framework convention on climate change, 1997.

[65] UN. United nations framework convention on climate change. adoption of the paris agreement, 2015.

[66] UNECE Global technical regulation No. 15. Worldwide harmonized Light vehicles Test Procedure, 2014.

[67] M. Vint. 48v Mild Hybrid Vehicle Systems. In *SAE 2016 Hybrid and Electric Vehicle Technologies Symposium*, 2016.

[68] H. Wallentowitz, J.-W. Biermann, R. Bady, and C. Renner. Strukturvarianten von hybridantrieben. In *VDI-Tagung Hybridantriebe*, 1999.

[69] H. Wallentowitz and K. Reif. *Handbuch Kraftfahrzeugelektronik - Grundlagen - Komponenten - Systeme - Anwendungen*. Springer-Verlag, Berlin Heidelberg New York, 2. aufl. edition, 2010.

[70] J. Warnatz, U. Maas, and R. W. Dibble. *Verbrennung*. Springer-Verlag, Berlin Heidelberg New York, 3. aufl. edition, 2013.

[71] F. Winke. Untersuchung der Dynamikanforderungen an Verbrauchssimulationen von hybriden Antriebssträngen. Diplomarbeit, Universität Stuttgart, 2012.

[72] F. Winke, H.-J. Berner, and M. Bargende. Dynamische Simulation von Stadthybridfahrzeugen. *Motortechnische Zeitschrift*, 2013.

[73] F. Winke, H.-J. Berner, and M. Bargende. Dynamic simulation of hybrid powertrains using different combustion engine models. In *12th International Conference on Engines & Vehicles, Capri, Naples (Italy)*, 2015.

[74] G. Woschni. Die Berechnung der Wandwärme und der thermischen Belastung der Bauteile von Dieselmotoren. *Motortechnische Zeitschrift*, (31), 1970.

[75] F. Zacharias. *Analytische Darstellung der thermischen Eigenschaften von Verbrennungsgasen*. PhD thesis, Technische Universität Berlin, 1966.

Printed in the United States
By Bookmasters